◁ **Overleaf** A sunset illuminates lenticular clouds over the Southern Alps in New Zealand. The clouds form in the crests of waves that can develop in winds as they blow over hills and mountains (see *page 52*).

◀ **Freezing rain, or glaze,** forms a coating of ice on surfaces (see *page 12*). The weight of the ice after a heavy fall of freezing rain can snap branches off trees and cause wires to sag.

weather explained

explained

A Beginner's Guide to the Elements

DEREK ELSOM

A Henry Holt Reference Book
Henry Holt and Company
New York

Henry Holt and Company, Inc.
Publishers since 1866
115 West 18th Street
New York, New York 10011

Henry Holt® is a registered trademark
of Henry Holt and Company, Inc.

Published in Canada by Fitzhenry & Whiteside Ltd.,
195 Allstate Parkway, Markham, Ontario L3R 4T8.

Library of Congress Cataloging-in-Publication Data
Elsom, Derek M.
    Weather explained : a beginner's guide to the elements / Derek
Elsom.
        p. cm.—(Henry Holt reference book) (Your world explained series)
    Includes index.
    ISBN 0-8050-4875-8 (alk. paper)
    1. Weather. 2. Meteorology. I. Title. II. Series. III. Series: Your world
explained series.
QC981.E57 1997   97-11968
551.5—dc21   CIP

ISBN 0-8050-4875-8

Henry Holt books are available for special promotions
and premiums. For details contact: Director, Special Markets.

First Edition—1997

Editor: Katrina Maitland Smith
Designer: Steve Woosnam-Savage
Managing Editor: Kate Phelps
Art Director: Ralph Pitchford
Editorial Director: Cynthia O'Brien
Production: Janice Storr, Selby Sinton
Research: Lynda Wargen
Picture Research: Zilda Tandy, Elizabeth Loving

Printed and bound in Portugal by Printer Portuguesa
Originated in Singapore by Master Image
All first editions are printed on acid-free paper.∞

10 9 8 7 6 5 4 3 2 1

WARNING
Although weather watching is fun you must be careful when looking up at the sky. Never look directly at the Sun, even through sunglasses, because its light can seriously damage your eyes and even blind you.

Contents

WEATHER WORKINGS

What is weather?	8
How clouds form	10
Raindrops and frozen jewels	12
Why do winds blow?	14
Wind patterns	16
Weather fronts	18
The Earth's climate	20

WEATHER EXTREMES

Thunderstorms	22
Living with lightning	24
Downbursts and ice falls	26
Terrifying twisters	28
Tornado tales	30

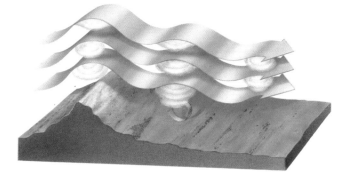

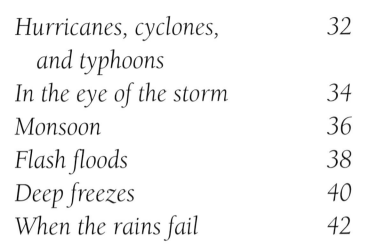

Hurricanes, cyclones, and typhoons   32

In the eye of the storm   34

Monsoon   36

Flash floods   38

Deep freezes   40

When the rains fail   42

THE WORLD'S CHANGING CLIMATE

Climate clues   54

Ice ages   56

Darkness and flickering light   58

El Niño   60

Polluting the atmosphere   62

The greenhouse effect   64

GLOSSARY   66

INDEX AND ACKNOWLEDGMENTS   68

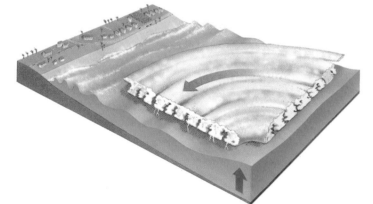

WEATHERWISE

Collecting information   44

Making weather forecasts   46

Using weather forecasts   48

Skywatching   50

Cloudscapes   52

WEATHER WORKINGS

WEATHER AFFECTS US ALL. IT AFFECTS WHAT CROPS WE GROW, how we build our homes, and what clothes we wear. Weather is dangerous when storms, floods, and droughts occur. Knowing why and how weather changes can be vital.

What is weather?
8–9

How clouds form
10–11

Raindrops and frozen jewels
12–13

Why do winds blow?
14–15

Wind patterns
16–17

Weather fronts
18–19

The Earth's climate
20–21

summer in northern hemisphere; winter in southern hemisphere

What is weather?

Weather is what is happening in the air outside at this moment. It is described by what you see, feel, and sometimes hear. You may see a bright, clear blue sky and feel very hot. Or you may see rain falling from dark gray clouds and feel slightly chilly. Within a short time, flashes of lightning and rumbles of thunder may follow.

In some parts of the world, people experience the same weather day after day for long periods. In other places, the weather changes from hour to hour and you can never be sure what to expect next.

▲ Satellite pictures of the Earth show whirls and bands of clouds bringing rain and stormy weather to parts of the Earth's surface.

The weather layer

There is an enormous variety of weather occurring on Earth at this moment. Some people are sweltering in record high temperatures while others are bitterly cold in temperatures below the freezing point (32°F). Many people are enjoying calm and pleasant weather but some are being battered by hurricane-force winds, torrential rainstorms, or golf-ball-sized hailstones.

Around the Earth is an envelope of gases called the atmosphere. It contains the air we breathe and protects us from harmful rays from the Sun and from space. Nearly all weather forms in the layer of the atmosphere that lies closest to the Earth's surface. Called the troposphere, this layer is surprisingly thin. If the Earth were reduced to the size of a large apple, the troposphere would be less than the thickness of the apple skin. It is about 8 miles deep midway between the poles and the equator, but is a little deeper at the equator itself (about 11 miles) and thinner at the poles (about 5 miles). Within this layer are to be found all the clouds, storms, wind, rain, hail, and snow that make up our weather.

spring in northern hemisphere;
fall in southern hemisphere

Earth's axis is
tilted 23.5° from
the vertical

height in miles

600

winter in northern
hemisphere; summer in
southern hemisphere

Sun
(not to scale)

Sun's rays

fall in northern hemisphere;
spring in southern hemisphere

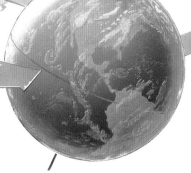

▲ The Earth spins on
an axis that is tilted from the
vertical. As the Earth orbits
the Sun, each hemisphere
(the half of the Earth above
or below the equator) takes
a turn at being tilted toward
the Sun—this is summer.
During summer the Sun is
higher in the sky and long
hours of sunshine bring
warmth. Winter occurs
when the hemisphere is tilted
away from the Sun.
In spring and fall, the
Sun lies over the
equator.

▼ Seen from space,
the movements and changing
shapes of clouds above the
Earth's surface show weather
forecasters where storms
are and how they are
developing.

**▶ The Earth's
atmosphere** is made
up of several layers. Nearly all
our weather happens within
the troposphere. Electrical
charges from thunderstorms in
the troposphere (see page 23)
sometimes shoot bright lights—
glowing blue jets and red sprites—
into the upper layers. Aurorae are
shimmering curtains of light caused
by bursts of energy from the Sun,
which cause gases in the upper
atmosphere to glow.

—exosphere

—high
aurora

300

—thermosphere

low
aurora

mesosphere
stratosphere
troposphere

50 sprite
30
8 jet

frontal depression
(see pages 18–19)

dissolving
frontal
depression

developing frontal
depression

low pressure zone
(see pages 14–15)

thunderstorms
(see pages 22–23)

high cirrus clouds
along jet stream
(see pages 18–19)

hurricane
(see pages 32–33)

clear skies in high
pressure zone
(see pages 14–15)

How clouds form

Clouds are the visible signs that the atmosphere contains water. Water is always present in air but it is normally invisible in the form of a gas, called water vapor. Clouds occur when the air is chilled until the water vapor changes (or condenses) into its liquid form, as water droplets. These droplets are so tiny that they remain suspended in the air, making the cloud. When it is very cold, the water vapor freezes straight into ice crystals.

The usual way in which air is chilled to form a cloud is by the air rising within the troposphere. In this part of the atmosphere, temperature decreases with height. The temperature at the top of the troposphere is well below freezing, at about −67°F.

Cloud types

Various shapes, sizes, and thicknesses of cloud are formed by the different forces that cause air to rise, cool, and produce the cloud. On sunny days, air rising from sun-warmed slopes and fields usually produces small, dome-shaped clouds called cumulus. If a cumulus continues to grow, it may become a giant cumulonimbus, which is also known as a thunderhead.

Stratus clouds are layer-shaped. Ragged, gray stratus forms when air is forced to rise as it passes over hills.

When a mass of warm air glides up over a mass of cold air, a number of different cloud types form along the boundary between the two air masses. These include rain-bearing clouds called nimbostratus, and altostratus clouds through which watery sunlight appears.

Fog is a cloud that rests on the ground. It is usually stratus and often forms during long, clear nights when the ground quickly loses its surface heat. The air near the ground is chilled until the water vapor in the air condenses to form tiny water droplets. A shallow layer of fog is produced. It is usually burned off the next morning by the Sun's warmth.

condensation level (height at which water vapor condenses into water droplets)

thermal

cloud forms as water vapor condenses into water droplets

cooler air

warmer a

◀ Cumulus clouds form over warm areas and drift away with the wind. They often develop during late morning and sometimes produce showers in the late afternoon.

▶ Clouds are named for their shape and height. In appearance, clouds can be cumulus (which is the Latin word for "heap"), stratus (the Latin for "layer"), and cirrus ("hooked" or feathery). They occur at different heights: low, medium (called alto), and high although cumulonimbus may extend from the bottom to the top of the troposphere. Nimbus is Latin for "cloud" but usually refers to a rain cloud.

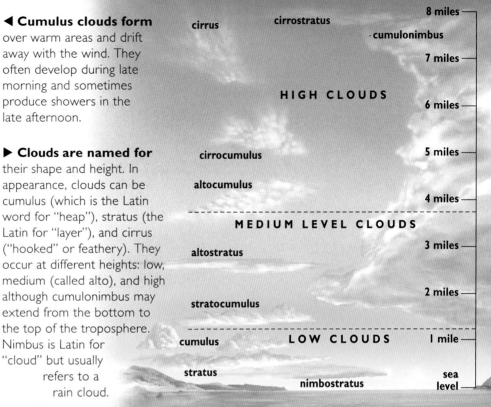

		8 miles
cirrus	cirrostratus	
		cumulonimbus
		7 miles
	HIGH CLOUDS	
		6 miles
		5 miles
cirrocumulus		
altocumulus		
		4 miles
	MEDIUM LEVEL CLOUDS	
altostratus		3 miles
		2 miles
stratocumulus		
	LOW CLOUDS	1 mile
cumulus		
stratus		sea level
	nimbostratus	

cloud drifts away as new cloud begins to form

A. warm ground

B. hills and mountains

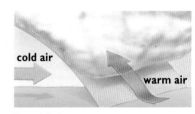

cold air

warm air

C. air from opposite directions

D. cold air mass

◀ On a sunny day, hillsides facing the Sun and areas of bare soil or rock warm more than the surrounding land. A large bubble of air, called a thermal, rises. The air cools as it rises and water vapor condenses into tiny cloud droplets. Wind drags the small cloud away and the process begins again.

▲ Clouds form from rising air when: **A** the Sun warms the ground and creates thermals; **B** air moves up over a hill or mountain; **C** air from opposite directions meets and is forced upward; **D** cold air flows under and lifts warmer, lighter air.

11

Raindrops and frozen jewels

The tiny water droplets and ice crystals that form clouds are far too small and light to reach the ground as rain or snow. Many millions of tiny cloud droplets must join together to create water droplets that are heavy enough to fall to the ground. Raindrops can grow to around one fifth of an inch in diameter. Because large raindrops fall faster, they often reach the ground first and mark the start of a rain shower. Similarly, vast numbers of ice crystals must be stuck together to build a snowflake large enough to reach the ground. Snowflakes are so delicate that they often bump and shatter as they fall. So a snowflake reaching the ground is actually a jumble of several snowflakes.

▲ Snowflakes
are formed from tiny ice crystals in a cloud, which join together to make delicate, six-sided (hexagonal) star shapes. Each beautiful snowflake is different from every other.

◄ Frost can form
on calm nights if the ground temperature is below the freezing point. Air that touches the ground may be chilled so that water vapor freezes into sparkling ice crystals. If the ground temperature is above the freezing point, a thin coating of water droplets, called dew, may form.

water droplets

larger droplets form

condensation level
(see *page 10*)

drizzle

rain

Precipitation

The air within a cloud is constantly moving, carrying water droplets and ice crystals with it. As they move, the droplets or crystals bump into one another and join together. Larger droplets or crystals sweep up smaller ones in their path, until they are too heavy to stay suspended in the air and they fall to the ground.

All forms of moisture that come from the atmosphere are known as precipitation. Rain and snow are the most common types, but there are many others. Some of these are shown in the picture above.

Some water droplets freeze onto ice crystals, making a clear ice pellet or a milky snow pellet.

When very cold rain touches ground that is below the freezing point, it produces a heavy and glassy coating of ice on surfaces, called freezing rain or glaze.

cumulonimbus cloud

▼ **Sleet** is a name that causes confusion. Americans describe it as frozen water droplets. These clear ice balls bounce on striking hard ground. To the British, sleet is either falling snow that is melting or a mixture of snow and rain.

water droplets and ice crystals

water droplets and ice crystals

hail

dry snow

wet snow

sleet

rain

◀ **Types of precipitation** depend on whether a cloud carries water droplets, ice crystals, or both. Shallow, low clouds contain only droplets, so produce drizzle or rain. Deeper clouds contain droplets and crystals, so produce rain, snow, or sleet. Hail forms in cumulonimbus (*see also page 27*).

Rainmaking

When there is a shortage of water for crops, scientists have attempted to make it rain or snow. A deep cloud, which could produce a lot of precipitation, normally needs a few very large cloud droplets or ice crystals to be present to encourage others to begin joining together to produce rain or snow. Ice crystals work best. Scientists can try to trick the cloud by "seeding" it with substances that look like large ice crystals. Chemicals such as dry ice, which is the smoky vapor used in theaters to give scenes a misty look, are sprayed into the cloud from airplanes.

This method has been tried in the Sierra Nevada in California and the Rocky Mountains in North America for many years. When successful, the snowfall—or rainfall if the snow melts before it reaches the ground— is increased by about one-sixth.

THE WATER CYCLE

Water in the atmosphere is part of a continual cycle, which is essential for life on Earth. Water enters the atmosphere as water vapor, taken up from rivers, lakes, and oceans as well as from plants and animals, in a process known as evaporation (**1**). As air rises and cools, the water vapor within it condenses into water droplets and clouds form (**2**). The clouds return the moisture to the Earth's surface as precipitation (**3**). The water runs off the land and seeps into the ground (**4**), making its way back to rivers and oceans.

Why do winds blow?

Air moves from place to place as wind. This happens because some parts of the Earth are warmed by the Sun more than others. Air over warmer ground will increase in temperature. As the air gets warmer, it expands (meaning that the same amount of air fills a larger space) and so gets lighter. Because the warmer air is lighter than the surrounding air, it rises, just like a hot-air balloon.

The rising air is replaced by air flowing in from surrounding areas, particularly from above cooler surfaces where the air is cooling, getting heavier, and sinking. This is what creates winds.

Winds are slowed in the lowest 3,300 feet of the atmosphere because they drag along the Earth's surface. They may be slowed by one-third over the sea and by two-thirds over rough land.

▲ **Winds can range from light** sea breezes to enormously powerful and destructive winds that whirl around hurricanes and tornadoes.

Winds and air pressure

The weight of air above the ground produces a force pressing down on the Earth's surface. This is called air pressure. In places where the air is rising, there is less pressure pushing down on the surface. This creates an area of lower pressure. Where air is sinking toward the surface, it produces an area of higher pressure.

Weather forecasters draw maps of air pressure. Lines called isobars are drawn joining up places that have the same air pressure. Where the isobars make circular patterns, these are the areas of low and high pressure, known as lows and highs. From these maps (*see also page 47*),

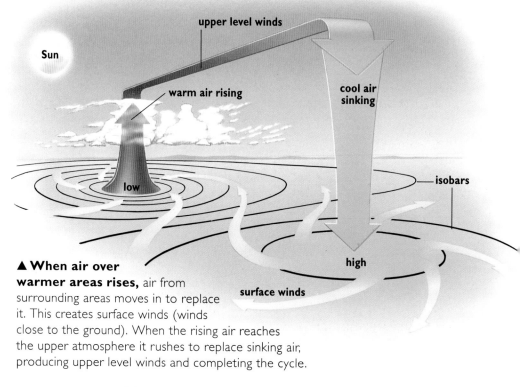

▲ **When air over warmer areas rises,** air from surrounding areas moves in to replace it. This creates surface winds (winds close to the ground). When the rising air reaches the upper atmosphere it rushes to replace sinking air, producing upper level winds and completing the cycle.

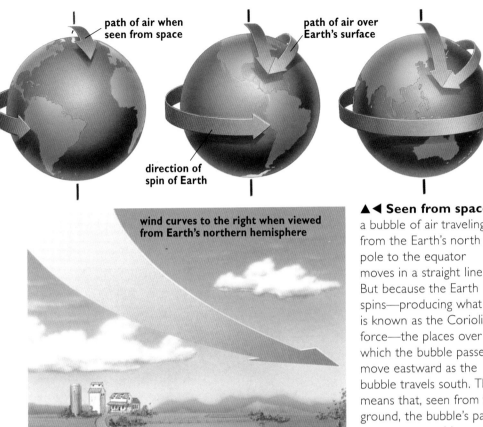

path of air when seen from space

path of air over Earth's surface

direction of spin of Earth

wind curves to the right when viewed from Earth's northern hemisphere

▲◄ **Seen from space,** a bubble of air traveling from the Earth's north pole to the equator moves in a straight line. But because the Earth spins—producing what is known as the Coriolis force—the places over which the bubble passes move eastward as the bubble travels south. This means that, seen from the ground, the bubble's path curves to one side.

forecasters can determine which way the winds are blowing. Winds blow from high to low pressure. The greater the difference in air pressure between highs and lows, the stronger the winds will be.

Winds do not blow in a straight line from highs to lows. This is because the Earth spins. The spin produces the Coriolis force (*see above*), which throws winds sideways from their path. In the northern hemisphere the wind is thrown to the right of its path; in the southern hemisphere, it curves to the left. The dragging effect of the Earth's surface also prevents winds from blowing in a straight line.

Lows and highs

Lows that have very low air pressure create strong winds. The rising air in lows often produces clouds and rain. These lows are the storms such as frontal depressions (*see pages 18–19*) and hurricanes (*pages 32–35*).

High pressure areas are also called anticyclones. Air within anticyclones is sinking and creates clear skies because it stops the air from rising and producing clouds and rain. The higher the air pressure in anticyclones, the drier and clearer the air. When anticyclones sit over a region for several days they can bring bitterly cold weather in winter and hot, sunny weather in summer.

(see pages 18–19); (pages 32–35)

THE BEAUFORT SCALE

In 1806, the British Admiral Sir Francis Beaufort (1774–1857) worked out a scale for estimating wind strength at sea by looking at the height and appearance of waves. The effects of winds on land were added later. Force 1 means winds of 1–3 mph; force 6, winds of 25–31 mph; and force 12, winds of 74 mph or more. Hurricane winds can reach more than 200 mph. If the Beaufort Scale included the incredible wind speeds in tornadoes it might reach force 29 or 30.

0	Calm	Smoke rises straight up
1	Light air	Smoke drifts
2	Light breeze	Wind felt on face
3	Gentle breeze	Leaves and twigs move
4	Moderate breeze	Flags flap
5	Fresh breeze	Small trees sway
6	Strong breeze	Large branches move
7	Near gale	Whole trees sway
8	Gale	Twigs break off trees
9	Strong gale	Branches break off trees
10	Storm	Trees uprooted
11	Violent storm	Widespread damage
12	Hurricane	Disaster

Wind patterns

Labels on the globe diagram (left):

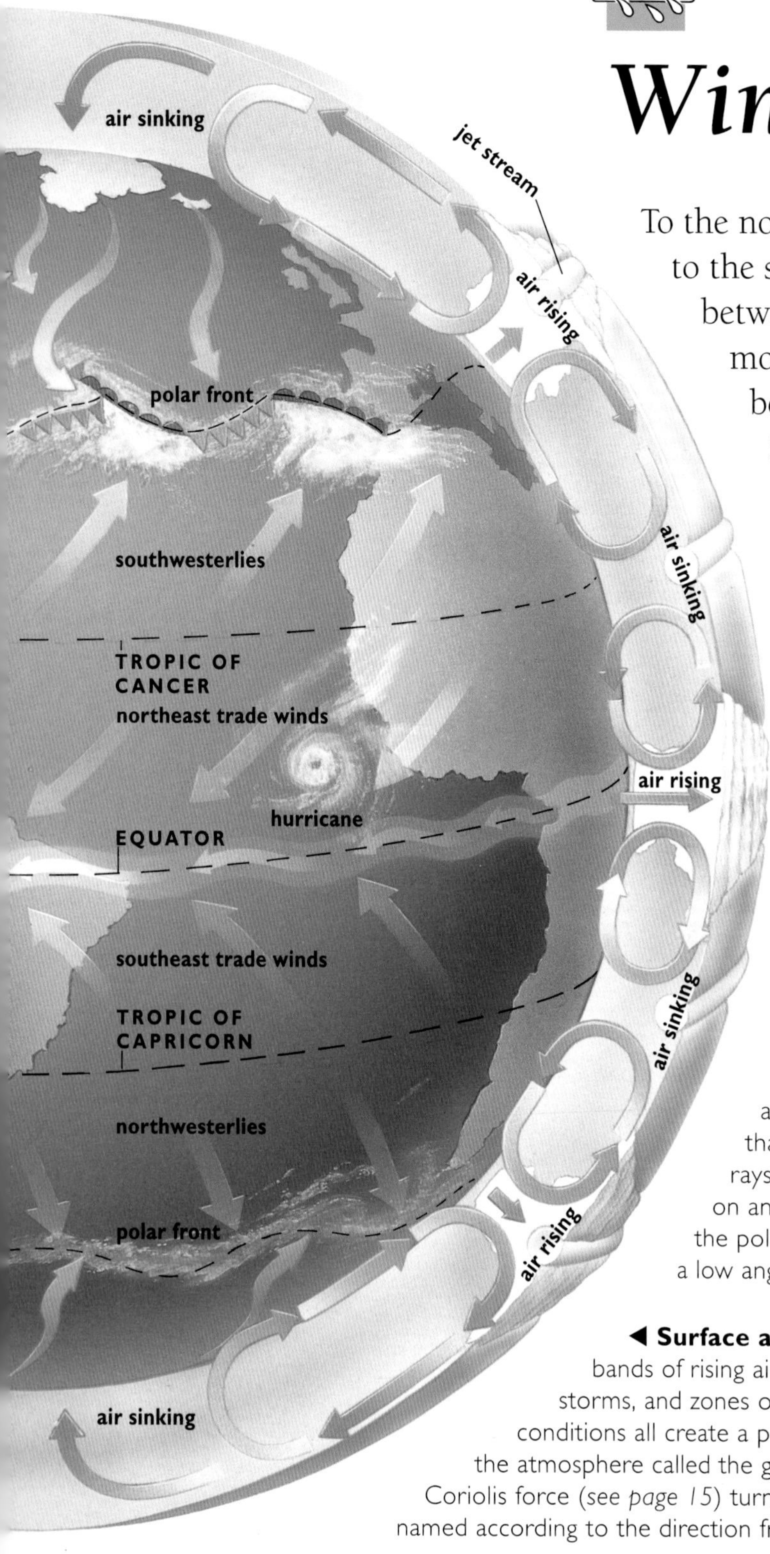

air sinking

jet stream

air rising

polar front

air sinking

southwesterlies

air sinking

TROPIC OF CANCER
northeast trade winds

air rising

EQUATOR

hurricane

southeast trade winds

air sinking

TROPIC OF CAPRICORN

northwesterlies

polar front

air rising

air sinking

To the north of the Earth's equator lies the Tropic of Cancer; to the south lies the Tropic of Capricorn (*see left*). The region between the two is known as the tropics and it receives far more heat from the Sun than areas outside it. This is because the Earth is ball shaped, which results in the Sun's heat being concentrated in the tropics. If the atmosphere did not exist, the tropics would get hotter and hotter while other parts would get colder. The atmosphere stops this happening by moving both the hotter and colder air around the planet.

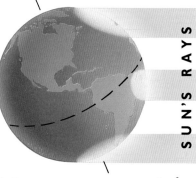

SUN'S RAYS

▲ **Because the Earth** is ball shaped, the tropics are heated more strongly than the poles. The Sun's rays hit the tropics nearly full on and are concentrated. Near the poles the Sun's rays strike at a low angle and are spread out.

◄ **Surface and upper winds,** bands of rising air producing clouds and storms, and zones of sinking air creating drier conditions all create a pattern of movement in the atmosphere called the global circulation. The Coriolis force (*see page 15*) turns the winds, which are named according to the direction from which they blow.

Global winds

The equator is the driving force behind the global wind pattern. Here, intense heating by the Sun creates giant and violent cumulonimbus clouds and heavy rainfall. The fast-rising equatorial air draws in strong, steady winds called the trade winds from each hemisphere. Hurricanes may form in the trades.

At the Earth's poles, with their vast areas of snow and ice, cold air sinks and spreads outward. When it reaches the Earth's middle latitudes (midway between the equator and the poles) it meets warmer winds along a boundary called the polar front. As the two winds collide they produce the storms called frontal depressions (*see pages 18–19*).

DAY

warm air rising

cool air sinking

sea breeze

NIGHT

cool air sinking

land breeze

warm air rising

◄ **On sunny days, the land** warms faster than the sea. The warmed air above the land rises and draws in cooler sea air to replace it, called the sea breeze. At a height of about 3,300 feet, the air above the land blows out to sea. During late evening, the flows are reversed because the sea keeps its warmth longer than the land. Air flowing away from the shore is called a land breeze.

▼ **Winds on valley slopes** change direction from day to night. During daytime, the sunny slopes warm quicker than the valley bottom, causing the air to rise. Air from the valley below moves up to replace it, creating a valley breeze. At night, the upper slopes cool faster than the valley floor and the chilled air sinks, producing a mountain breeze. The air may be so chilled that fog forms (*see also page 10*).

Local winds

Mountains, valleys, and coasts often create their own local winds to which people living in the area give special names. Some local winds can create major problems.

The Santa Ana is the name given to a warm, dry wind that blows through mountain canyons into southern California. Like the chinook (*see below right*), it gains warmth because it is compressed (so that the same amount of air is squeezed into a smaller space) as it descends from the lower air pressure of the mountains to the higher pressure in the valleys. The Santa Ana can dry out vegetation and often leads to major brush fires with the wind-swept flames destroying hundreds of homes.

▶ **A wind blowing up a mountain** produces clouds and rain or snow. Having lost its moisture as precipitation, the dry wind is warmed by compression as it rushes down the other side of the mountain. This hot, dry wind, called a chinook in the Rockies and a föhn in the Alps in Europe, melts the snow in spring.

air rises, forming clouds and precipitation

cool air carrying water vapor

warm, dry air

Weather fronts

When a mass of cold air meets a mass of warm air, the two do not simply mix. A boundary, called a front, remains between them. The leading edge of a warm air mass is called a warm front; the leading edge of a cold air mass is a cold front.

Frontal depressions bringing clouds and rain form at the polar front, where cold polar air and warm tropical air clash. The warm air pushes into the polar front making it bulge and the air begins to spiral slowly around a low pressure center (called a depression). At the warm front, the warm air glides up over the polar air producing clouds and drizzle. A cold front moves in behind pushing under the warm air and creating more clouds. When the cold front catches up with the warm front, all the warm air is lifted clear off the ground and the frontal depression dissolves (or occludes).

▲ **As a cold front passes,** nimbostratus and cumulonimbus clouds give way to clear skies and occasional cumulus clouds.

Rocky Mountains

trough

jet stream

developing frontal depression

fully developed frontal depression

WARM TROPICAL AIR

The jet stream

There are particular positions along the polar front where depressions tend to form. These positions depend on what is happening in the atmosphere above.

At about 30,000 feet are found fast-flowing ribbons of air called jet streams. They circle the Earth from west to east and are present in both hemispheres. Jet streams get their energy from the difference in temperature between the cold and warm air across the polar front. The greater the difference, the stronger will be the jet, which reaches speeds of more than 100 mph.

Each jet stream moves like a wriggling snake in the atmosphere. Sometimes it blows in an almost straight line but at other times it may have four or five bends in it. Where the jet stream loops away from the pole, it makes a shape called a trough, which is where depressions usually form. It is here that surface air is most easily drawn upward creating low air pressure at the ground and providing the pivot around which the air begins to slowly whirl. A trough is often found near the east coast of North America.

cumulonimbus

cirrus

cirrostratus

warm air

cold and warm
fronts merge

warm air

altostratus

cumulus
clouds

cold air

heavy rain

nimbostratus

rain

cold air

cold front

warm wind

warm front

cool wind

surface of Earth

◀ **The first signs**
of an approaching
depression are the feathery
cirrus clouds, which then spread
out to form cirrostratus. Altostratus
and nimbostratus follow, as the point
where the warm front reaches the ground
comes closer. The warm sector between the
fronts may have a few stratus or stratocumulus
clouds. The cold front brings cumulonimbus
and heavy rainfall.

COLD POLAR AIR

occluding depression

ridge

◀ **Each jet stream**
steers the frontal depressions
beneath it. If weather forecasters
know the position of a jet they can
decide whether a depression below it will
affect a particular area or not. Unfortunately,
jets change their positions from day to day.
      The height of the Rocky Mountains causes
the jet stream above to swing southward, often
producing a trough (or area of cold air and lower
pressure) to the east. A ridge is an area of warm
air and higher pressure.

Frontal depressions

Several days pass from the first
signs of a frontal depression forming
to when it fades away. During that
time the depression may travel
thousands of miles and bring much
changeable and stormy weather to
areas over which it passes.
      A frontal depression
in its fully developed
stage—when its clouds appear
comma-shaped from a weather
satellite—is a large storm sometimes
1,000 miles across. The reason for
its large size is that sequences of
clouds form along the fronts and
these fronts lie at a surprisingly
shallow angle to the ground (*see
above*). This means that the clouds
hugging the warm front at the top of
the troposphere lie perhaps as much
as 600 miles ahead of the clouds at
the bottom of the front. The angle
of the cold front is a little steeper.

The Earth's climate

The yearly pattern of weather in one place is called its climate. Climate is influenced by latitude (this is the measure of how far north or south of the equator somewhere is). Hotter climates are found in the tropics, colder ones around the poles, and warm (or temperate) climates in between. The global arrangement of land, sea, ocean currents, and winds all affect climate.

Places with similar climates can be grouped into a climate type even though they may be far apart. Boundaries between types are only approximate and they shift over time.

Mediterranean: mild winters with light rains; warm or hot dry summers

City climates

A map of global climate types ignores small differences between places. However, cities add their own distinctive local changes to temperature and precipitation.

Cities release a lot of heat from homes, factories, and motor vehicles, which warms the air above. During the day buildings, roads, and pavements absorb heat from the Sun and release it at night. All this heat results in cities—especially city centers—being much warmer than the surrounding countryside at night. Sometimes the difference can be as much as 18°F.

City warmth and the height of buildings helps air to rise. This can increase the depth of clouds over cities and trigger thunderstorms. As a result, rainfall may be one tenth higher over cities.

Tropical forest: hot and wet all year; heavy showers most days

Tropical grassland: hot all year; wet and dry seasons due to monsoon (see pages 36–37)

▶ **Climate is grouped into** different types depending on temperature and precipitation. The names of the climate types sometimes refer to the typical vegetation that is present because it has adapted to a particular climate over a long time.

Temperate: cool or mild winters; warm or hot summers; rain all year or winter rains

Polar and tundra: very cold and long winters; dry all year; soil frozen all year or thaws in summer

Cold forest: cold winters; warm summers; light rains

Desert: hot and dry all year; hot days, cold nights

Mountains: cold high up, warmer lower down; wet on windward slopes (slopes facing the wind), dry on leeward slopes (slopes sheltered from the wind)

Dry grassland: cold, wet winters; hot, dry summers

▶ **Temperature decreases with height** by 3.6°F every 1,000 feet. Mountains can have temperate forests at the foot (or rain forests if they are in the tropics) and cold, snowy conditions at the top. Many mountains have a wet and a dry side because winds give up their moisture as precipitation on the side facing the wind.

WEATHER EXTREMES

EVERY NOW AND THEN SEVERE WEATHER OCCURS. SOMETIMES IT IS SLOW to develop. Days get a little hotter and drier until a heatwave and drought arrive. On other occasions, extremes arrive suddenly and unexpectedly. Violent storms bring floods and damaging winds.

Thunderstorms
22–23

Living with lightning
24–25

Downbursts and ice falls
26–27

Terrifying twisters
28–29

Tornado tales
30–31

Hurricanes, cyclones, and typhoons
32–33

In the eye of the storm
34–35

Monsoon
36–37

Flash floods
38–39

Deep freezes
40–41

When the rains fail
42–43

Thunderstorms

Towering, anvil-topped thunderstorms bring many weather threats. They can cause heavy rain and flash floods, damaging hailstones the size of golf balls, sudden powerful bursts of wind, and terrifying tornadoes. Yet what thunderstorms are best known for are lightning and thunder.

At this very moment there are about 2,000 thunderstorms raging across the Earth, each generating as many as 100 flashes of lightning a minute. Above them, brief bursts of glowing lights called jets and sprites shoot upward reaching many times the height of the thunderstorms themselves.

▼ **A thunderstorm forms from a small cumulus** when there is a strong upcurrent of warm, moist air (1). Precipitation falls dragging air down to create downcurrents: a thunderstorm has been born (2). The thunderstorm dies when the upcurrent of warm, moist air stops (3).

Inside a thunderstorm

Thunderstorms can form on hot days when moist air close to the ground is heated and rises quickly. Strong upward and downward rushes of air (called upcurrents and downcurrents) within the cloud sweep ice crystals, water droplets, and ice pellets past and into one another. This creates static electricity, which begins to build up. Negative electrical charges collect in the middle and lower parts of the cloud and positive charges gather toward the top. The difference between the charges builds up until it is so great that a massive spark is released as the charges even themselves out again.

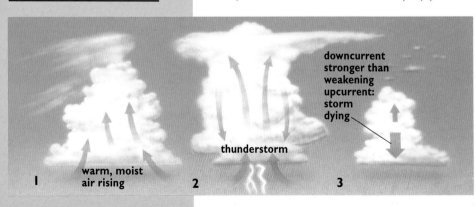

warm, moist air rising

1

thunderstorm

2

downcurrent stronger than weakening upcurrent: storm dying

3

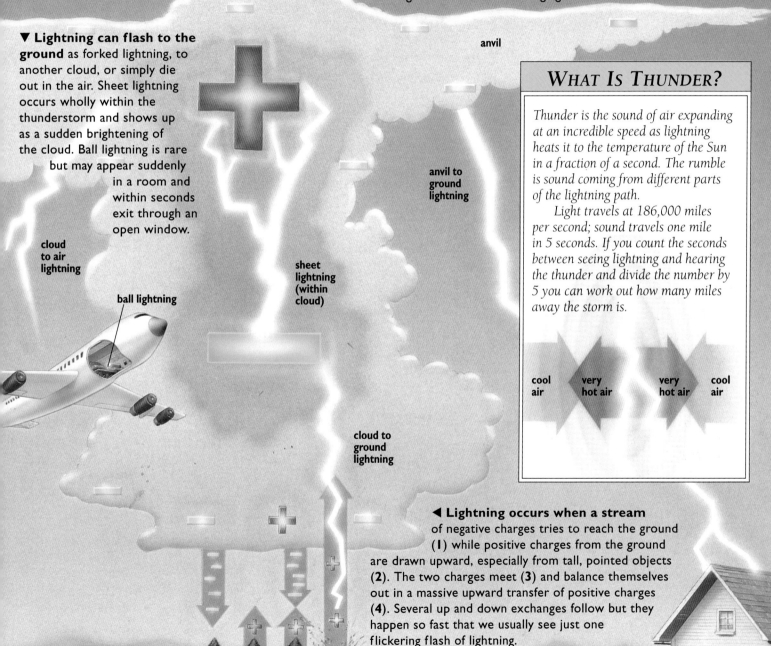

sprite

jet

◄ Glowing lights leap high above thunderstorms. Narrow blue jets shoot up to 20 miles above the Earth. Sprites reach 57 miles high and look like giant jellyfish, usually red but sometimes with green-blue tentacles hanging down.

anvil

▼ Lightning can flash to the ground as forked lightning, to another cloud, or simply die out in the air. Sheet lightning occurs wholly within the thunderstorm and shows up as a sudden brightening of the cloud. Ball lightning is rare but may appear suddenly in a room and within seconds exit through an open window.

anvil to ground lightning

cloud to air lightning

ball lightning

sheet lightning (within cloud)

cloud to ground lightning

WHAT IS THUNDER?

Thunder is the sound of air expanding at an incredible speed as lightning heats it to the temperature of the Sun in a fraction of a second. The rumble is sound coming from different parts of the lightning path.

Light travels at 186,000 miles per second; sound travels one mile in 5 seconds. If you count the seconds between seeing lightning and hearing the thunder and divide the number by 5 you can work out how many miles away the storm is.

cool air   very hot air   very hot air   cool air

◄ Lightning occurs when a stream of negative charges tries to reach the ground (**1**) while positive charges from the ground are drawn upward, especially from tall, pointed objects (**2**). The two charges meet (**3**) and balance themselves out in a massive upward transfer of positive charges (**4**). Several up and down exchanges follow but they happen so fast that we usually see just one flickering flash of lightning.

1   2   3   4

ball lightning

Living with lightning

Fewer people are killed by lightning now than years ago when larger numbers of people worked outdoors on farms. Today, more people live and work in towns and cities surrounded by buildings which attract the strikes instead. High buildings are protected from damage by lightning rods. These provide a safe route for lightning to reach the ground.

People are much more aware of the threat of lightning today and know what action to take. Better weather forecasts have helped. Some golf courses and schools have their own detectors to warn of approaching thunderstorms.

▲ If your hair stands on end, move away quickly and find shelter since lightning may strike. Positive electrical charges are being drawn from the ground to make a path for lightning.

Lightning horror stories

Not all people who receive an electrical shock from lightning are struck directly. Someone standing near a tree that is struck by lightning may suffer a side flash. This happens because lightning always seeks the easiest route to the ground. It jumps from the tree to a nearby person because people offer less resistance to electricity than a tree.

Entire herds of cattle and teams of people playing sports can be toppled by one lightning strike. When lightning strikes the ground or an object, the electrical charge spreads outward along the ground

from the point of impact like ripples from a stone dropped in a pond. A person or animal standing near the strike receives a shock which passes up one leg and down the other. It is enough to throw a person to the ground but seldom causes serious injury. The wider

◄ Lightning can cause dramatic damage. The sap beneath tree bark and the water soaked into stone or brick can be turned to vapor in a fraction of a second. This causes violent explosions that split trees and destroy walls. A schoolgirl died in England in 1974 when she was hit on the head by bark from an exploding oak tree.

apart the legs, the larger is the shock. Cows and sheep have their legs wider apart so they suffer larger shocks, which often kill them.

Large numbers of fish may be killed if lightning strikes a lake because the electrical charges spread through the water. Swimmers must get out of water promptly if a thunderstorm is approaching.

24

◀ This lightning strike
to a 65-foot tree shows two upward streamers that did not make contact with the downward flow from the cloud. One is from near the top of the tree; the other is from a television antenna.

25

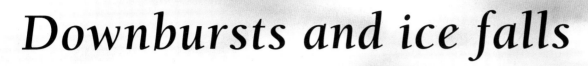

Downbursts and ice falls

Airplanes avoid flying through thunderstorms. Rapidly changing wind directions and speeds, called wind shear, make flights uncomfortable for passengers and can cause damage to the structure of airplanes. But the sky beneath a thunderstorm is especially dangerous. Powerful downcurrents can surge with such force that an airplane is slammed to the ground.

Strong thunderstorm upcurrents create another weather hazard: hail. The more powerful the upcurrents, the larger the hailstone grows before it falls. A 100 mph upcurrent can form a hailstone the size of a tennis ball.

▲ **Computers are now being used to train** airline pilots to cope with the sudden, powerful downward surge of air in a microburst.

Downbursts and microbursts

A downburst forms when very cold air carrying lots of heavy rain and snow plummets from the sky. It is like an enormous bucket of cold water being thrown onto the ground. The water hits the ground and bounces upward in all directions.

A wet downburst can be seen because of the precipitation it carries. In a dry downburst, the precipitation dries up before reaching the ground. All that is seen of a dry downburst is the dust raised by the powerful gusts of wind striking land.

When a downburst is less than 2½ miles across it is called a microburst. Winds can reach 168 mph in a severe microburst,

flight path

air surging downward

movement of air after striking ground

runway

▲ **An airplane** flying through a microburst first experiences a headwind (wind blowing from in front) and uplift. This suddenly switches to a tailwind (wind blowing from behind) and downcurrent. The airplane loses uplift and the pilot has only seconds to adjust the controls. The airplane may fail to reach the runway.

uprooting trees, damaging buildings, and causing airplanes to drop from the sky. In July 1994, as an airplane approached Charlotte Airport in North Carolina, a 40 mph headwind suddenly reversed and slowed to 30 mph. The airplane faltered and crashed killing 37 passengers and injuring 21 others. Eyewitnesses said the airplane just fell out of the sky.

To try to overcome the problem, airline pilots are now trained using virtual reality computers that imitate what happens to an airplane in a microburst. American airports also have special equipment to measure the winds so that pilots can be warned of a coming microburst.

ANTI-HAIL DEVICES

Several centuries ago, it was common practice in Europe to ring church bells when a hailstorm threatened. Farmers believed that the sound waves would break up large hailstones and save crops from the serious damage that they cause. Later, rockets were fired into storms in the hope that the particles of smoke would encourage large numbers of small hailstones to form rather than small numbers of large ones. Success has been limited.

◀ **These farmers in France in** 1910 are using anti-hail cannons to try to protect their fruit crops.

Hailstorms

A hailstone begins as a small ice or snow pellet. The pellet grows in size as it is swept up and down within a thunderstorm gathering layers of ice (*see picture on page 13*). Most hail is pea-sized, but hailstones the size of coconuts have fallen. When grapefruit-sized hail fell on students playing football in Colorado in 1979 nobody was hurt because they were all wearing helmets.

Hailstorms cause hundreds of millions of dollars of damage each year to crops in the Great Plains of the United States where farmers fear the "white plague." In China, India, and Bangladesh hailstorms kill and injure hundreds of people every few years.

▶▼ **Slicing through a hailstone** reveals onionlike layers, which record its up-and-down movements. In the cold, upper cloud levels, the hailstone collects water droplets which freeze quickly, trapping air bubbles and producing a layer of milky-white ice. At lower, warmer levels, freezing occurs more slowly, forming clear ice. Hailstones are usually ball shaped (*below*) but spiky chunks, plates, and fried-egg shapes have fallen.

▲ **Hailstones can break windows** and streetlights, dent cars and airplanes, batter roofs, and cause injuries in just a few minutes as happened in Munich, Germany, in 1984 when tennis-ball-sized hail fell.

Terrifying twisters

A tornado, or twister, is a powerful, funnel-shaped whirlwind of spinning, rising air. It is less than half a mile wide and sometimes only 50 yards across. It can leave behind a trail of incredible destruction as it roars across the landscape.

A tornado happens during a severe thunderstorm. It hangs beneath the cloud base of the storm and can take the shape of a slender, twisted rope, an ice-cream cone, or even a broad pillar. A very large tornado may have several small tornadoes rotating around it.

If a strong tornado cuts through a town, it can tear off roofs, flatten walls, uproot or snap trees, twist metal gates, smash sheds and mobile homes into splinters, topple trucks, and roll cars. It can pick up all sorts of debris and objects and carry them off for a considerable distance.

▲ **Wind speeds in very weak tornadoes** may be only about 40 mph but in violent tornadoes they can reach 300 mph. Such winds bring massive destruction, and can leave hundreds of people injured and homeless.

▶ **A tornado forms** from a particularly strong thunderstorm in which the winds high up in the storm cloud are blowing much faster than winds lower down and from a very different direction. This makes the thunderstorm rotate slowly, with the spin being strongest in the center. Like a spinning ice skater with arms pulled in and stretched upward, the center of the storm rotates faster. The spinning column of air eventually works its way down the storm's upcurrent toward the ground and appears as a funnel cloud.

rising air creates bulge, or dome, in top of cloud

front anvil

rear anvil

mamma (or mammatus) (see page 53)

storm cloud slowly rotates counter-clockwise

spin becomes concentrated in center of cloud

cumulus clouds attached to storm

CUMULONIMBUS

rain

hail

path of tornado

tornado

rain-free base of storm

Tornadoes and dust devils

A tornado is dangerous because its swirling winds may reach speeds of up to 300 mph. Generally, however, wind speeds in many tornadoes do not reach 100 mph.

The air in a tornado rises very rapidly. Heavy objects are lifted and lighter objects are sucked high into the air. Damage occurs only in the narrow strip where the tornado is in contact with the ground. Here, buildings may be destroyed completely, yet those 50 yards outside the tornado track may be untouched, or damaged only slightly by the debris being hurled aside.

Just one in 1,000 thunderstorms produces a tornado. Few tornadoes last for more than 15 minutes and

As a tornado scours the ground, it lifts and swirls sand, soil, and debris into the air. This can mask the lower part of the funnel and sometimes causes the tornado to change color.

many exist for only a few minutes. A path of tornado damage that covers hundreds of miles has usually been caused by a family of tornadoes, where the parent thunderstorm produces a new tornado after a previous one dies out.

Sometimes tornadoes are confused with dust devils. Dust devils are small whirlwinds that often form over plowed fields on hot, sunny days as warm air rises and is sent spinning by local winds. These fair weather whirlwinds may be only a few feet across and are not attached to a storm cloud. They are made visible by the dust, soil, and straw they lift from the ground. Dust devils seldom cause serious damage.

Storm chasers
Scientists are getting better at forecasting when and where a violent tornado may form. This knowledge is helping save lives. In the 1970s, tornadoes killed about 100 people each year in the United States. By the 1990s, this number had halved. Today, warnings of between 10 and 20 minutes are being given to towns across the country before a tornado strikes.

More information about tornadoes would further improve warnings. Since the 1980s, people known as storm chasers have tried to place weather instruments in the path of a tornado or to probe them at a safe distance using a special radar to find out more about what happens in the heart of a tornado.

29

Tornado tales

The power of a tornado can produce some strange effects. Stalks of straw have been driven into a telegraph pole, planks of wood shot through a barn door, and a playing card embedded on its edge more than an inch into a wooden door. Bark has been stripped from trees and feathers plucked from chickens. The amount of feathers lost by chickens was once suggested as a way of estimating the strength of a tornado's winds.

Trains have been lifted, turned around, and dropped onto the track facing the other way. Heavy refrigerators have been carried hundreds of yards while lighter objects have been carried for tens of miles.

▲ ▶ A waterspout can suck up objects from the sea or beach and carry them long distances before dropping them inland.

SEA

▲ Waterspouts are like tornadoes but over water and are usually much weaker. They form from fast-rising cumulus as well as thunderstorms. Sailors in past centuries thought that waterspouts were sea monsters and fired cannons at them.

Freak falls
A tornado passing over a pond or river can suck up the contents like a huge vacuum cleaner. Hundreds of small frogs, toads, tadpoles, fish, and weeds may be carried along for many miles until the tornado weakens and the objects fall to the ground or are flung to the side.

A nine-year-old girl living on a farm in England in 1932 suffered the strange effects of one tornado. She was out walking when a storm broke. Rain fell, but it felt soft and heavy. She shook her head and tiny frogs dropped to the ground. Her dog went berserk as frogs fell and tangled in its hair. Cows stampeded and the terrified girl ran home to her parents. They never believed her story.

Sometimes the tornado or waterspout that caused the freak falls is seen. Often, however, eyewitnesses are left baffled. In the past these strange falls were thought to be signs that something bad would happen.

▶ **Tornadoes occur so often in the Midwest** that it is known as Tornado Alley. Typically there are 1,000 tornadoes in the United States each year. Tornadoes occur in other parts of the world, including the United Kingdom, which has 15–30 small tornadoes each year. In 1989, the world's worst tornado disaster killed 1,300 people in Bangladesh.

▫	1
▨	3
▨	5
▨	7
▪	9

average number of tornadoes per 10,000 square miles per year in the U.S.A.

storm cloud rotates counter-clockwise with spin strongest in center

path of waterspout, which becomes a tornado over land

rain

objects dropped on land to one side of waterspout's path

waterspout

LAND

objects sucked up from sea or beach

In the path of a tornado

Being caught in a tornado can be very scary but most people survive, especially if they take shelter.

In 1995 a baby boy was plucked from his cot and carried away from his destroyed home in Des Arc, Arkansas, by a tornado. He was found safe in a ditch half a mile away, muddy and with just a few scratches and bruises.

In 1992, a young girl escaped unhurt after being carried almost two miles by a tornado near Shanghai, China. She was set down in a treetop. Trying to flee from a tornado in a car is not a good idea. A tornado is too fast and the direction it travels in is too unpredictable for drivers to know where to go to avoid it. This was made clear when a tornado with 200 mph winds struck Witchita Falls, Texas, in 1979. As it approached the city, some people jumped into their cars and tried to flee the tornado by driving away from it. However, 26 out of the 43 people killed and 30 out of the 59 people with serious injuries were in cars. Most of the victims' homes were left undamaged by the tornado's powerful winds.

TORNADO ALERT

In the United States a "tornado watch" is started for a region when conditions are likely to produce tornadoes in the next few hours. Radar and volunteer tornado spotters keep close watch. If a tornado is spotted, a "tornado warning" is sent out. Sirens may be sounded.

What to do:
1 *Do not try to escape in a car.*
2 *Leave a mobile home.*
3 *Seek shelter in a storm cellar or strong building.*
4 *Keep away from all windows and doors.*
5 *If there is no cellar, stay on the ground floor in a small inside room or under the stairs.*
6 *Cover your head in case of flying debris.*

▶ **TOTO** (the Totable Tornado Observatory) was built to collect information from inside tornadoes.

Hurricanes, cyclones, and typhoons

Hurricanes are monster tropical storms between 250 and 500 miles across. Although their wind speeds tend to be slightly lower than the worst tornadoes, their immense size means that they affect far more people.

Hurricanes are called typhoons in the northwestern Pacific Ocean and cyclones in the Indian Ocean and around Australia. The general name for these storms is a tropical cyclone. They begin as a cluster of thunderstorms which then spiral and grow into a giant storm. Only when the winds reach 74 mph or more is the storm called a hurricane.

path of hurricane

dry air sinking

cloud-free eye

strong upcurrents

direction of rotation of hurricane

low level winds flow inward

Hurricane dangers

The worst hurricanes have winds greater than 155 mph, even reaching 200 mph. They create giant waves as high as 25 feet that smash against the coast. Six to 12 inches of rainfall can be dumped by a hurricane in just a few hours, causing massive floods. When Hurricane Gilbert struck Mexico in 1987, five buses were swept away by raging flood waters with the loss of 140 lives.

The spiralling bands of thunderstorms that form the hurricane can also produce tornadoes to add to the hurricane's power of destruction. Most hurricanes produce between 10 and 20 tornadoes, but Hurricane Beulah in 1967 generated 115 twisters!

Calm at the center

The most remarkable part of a hurricane is the eye at its center. As the eye passes over a location it brings a spell of calm conditions. The smaller and more circular the eye, the more severe the hurricane. Hurricane Gilbert, one of the most violent hurricanes this century, had an eye only 8 miles wide at one point. Hurricanes move slowly—at around 15 mph—so the eye offers about half an hour of peace.

A heat engine

Hurricanes are one of the ways in which the atmosphere moves heat away from the tropics and so stops this region of the Earth from overheating (*see page 16*). As water evaporates from the warm tropical sea into the air, it carries heat with it. The water vapor rises in the air and condenses to form clouds. This releases the heat, which warms the air further. The warm air continues to rise, helping the clouds to grow into huge thunderstorms.

◄ The sun's heating of the tropical sea causes air to rise and produce tall clouds, especially cumulonimbus. In the center of the storm, the low air pressure at the water's surface draws in more air from the surrounding area. Air spirals in, is warmed, and forms more clouds, which gather in distinct spiral bands. The thunderstorms are joined together at their tops by the cirrostratus anvils. The sky is clear within the eye because some air is sinking from above the storm's center and is stopping clouds from growing there.

► Nine out of 10 hurricane deaths are due to drowning. Forty million people live along the Atlantic and Gulf coasts of the United States and much of the land is less than 10 feet high. There is little to stop the giant foam-tipped waves of a storm surge from sweeping ashore.

upper level winds flow outward

spiral of thunderstorms

spiral bands of rain

trade winds are drawn into storm

STORM SURGES

Far out at sea, the winds of a hurricane, together with the low air pressure at the water's surface, raise the sea level by about 3 feet. This creates a vast dome of water. As the hurricane approaches a shallow coastline, this mound of water piles up against the coast and the powerful winds continue to increase the wave height. If this happens at high tide, a destructive surge 10 to 20 feet high sweeps ashore. It rips boats from their moorings and batters buildings. In the United States, building codes try to ensure that properties near the coast are constructed to withstand the force of storm surges, are placed well back from the shore, and are built on raised land.

storm surge

rotation of hurricane

raised dome of water beneath eye

low pressure in eye

▲ Seen from a satellite, a hurricane appears as a giant cloudy disk with a central cloud-free eye. Satellites are vital for tracking the development and movement of these deadly storms.

▼ Tropical cyclones are born only where the seas are warmer than 80°F. They do not form at the equator because there is no Coriolis force (see *page 15*) at the equator to cause the winds to curve and draw the thunderstorms into spiralling bands. Tropical cyclones drift with the trade winds and eventually curve poleward, leaving the warm seas and dying out.

In the eye of the storm

Hurricanes are constantly tracked by satellite and from the ground because of the dangers they bring. Brave pilots fly airplanes carrying weather instruments through the dark and violent clouds that surround the calm eye of the hurricane. These flights gain vital information about whether a hurricane is growing stronger or weakening. Hurricane experts on board the airplanes also look for any sign that the storm is about to change direction. On some occasions, a hurricane heading straight toward the coast has suddenly turned away, preventing a possible disaster. But the opposite has happened, too, and a hurricane has struck land so unexpectedly that the people living there had little time to prepare.

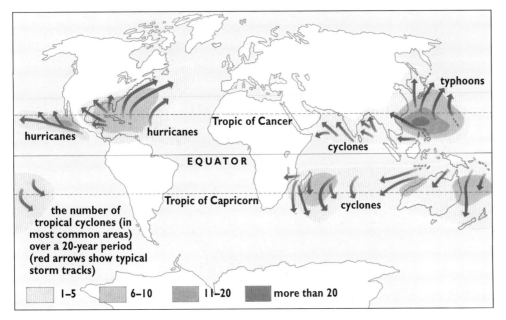

the number of tropical cyclones (in most common areas) over a 20-year period (red arrows show typical storm tracks)

1–5	6–10	11–20	more than 20	

Hurricane watches and warnings
When a hurricane is 36 hours away from reaching land, a "hurricane watch" is given out for a 300-mile stretch of the coastline. People living there must listen for a "hurricane warning," which tells them to evacuate. This will follow when the hurricane is 24 hours away and expected to strike a particular part of the coast.

The closer the hurricane is to the coast, the more accurate the warning of where it will hit land. But the warning has to be given at least a day in advance because the evacuation of thousands of people can take 12 to

Scale 1
minimal damage

Scale 2
moderate damage

Scale 3
extensive damage

Scale 4
extreme damage

Scale 5
catastrophic
damage

◄ **Each hurricane is given** a number from 1 to 5 to indicate how damaging it may be. It helps when deciding whether to board up windows, seek the safety of a storm shelter, or to evacuate the area. Class 5 hurricanes are the ones people fear most, such as Gilbert (*see also page 32*). Fortunately these are very rare.

▲ **When Hurricane Andrew hit Florida** in August 1992, it brought 16-feet-high storm surges and winds gusting to 168 mph. Yachts and fishing boats were washed far inland.

▼ **In 1970 a cyclone in Bangladesh** killed half a million people. Since then, hundreds of massive concrete shelters have been built to withstand cyclone storm surges.

24 hours. Because warnings have to be given so far ahead, some will turn out to be wrong. The National Hurricane Center in Miami has the difficult task of forecasting where a hurricane will strike land. If the Center gets it right, it saves lives. But if it makes too many mistakes, some people will not believe the next warning, which may turn out to be right. Incorrect warnings also cost money. To board up homes, close down businesses, move ships and boats to safer berths, and to prepare for evacuation can cost about $50 million.

Naming hurricanes

Hurricanes are given names so that everyone knows which storm is being talked about in watches and warnings. The first hurricane each year is given a name beginning with A; the second begins with B, and so on. For many years, only women's names were used but men's names were added after 1979. The names are used again every six years except in the case of killer hurricanes—these names are never used again. Names that have been retired because of the destruction the hurricanes caused include Andrew (1992), Gilbert (1987), Hugo (1988), and Joan (1988).

35

Monsoon

Dark clouds, torrential rain, and flooded fields and towns come to mind when we think of a monsoon. But these are features only of the wet, summer part of a monsoon. There is a dry, winter monsoon season, too.

Monsoon comes from the Arabic word *mausim* which means season. It refers to the seasonal change in the direction of the winds, especially those along the shores of the Arabian Sea. Here, the winds blow from the southwest during the summer months then swing around to blow from the opposite direction—from the northeast—during the winter.

Knowing when the wind would change direction was important to the shipping trade between east Africa and India in earlier centuries. The Portuguese explorer, Vasco da Gama, became the first European to reach India by sea in 1498 after an Arab sailor told him this weather secret.

▲ Streets covered by water are accepted as a small price to pay for the arrival of the life-giving summer rains. It is only when the rains are much worse than usual that people cannot cope, such as in September 1988 when two-thirds of Bangladesh was flooded and 28 million people lost their homes.

SUMMER MONSOON

Himalayas

hot land mass creates low pressure

Monsoons in Folklore

The monsoon is deeply rooted in the lives of the people of Asia. It is a key part of their legends, music, poetry, and religious festivals. In Indian poetry, dark clouds mean happiness, while the winter sun brings misery. Indian paintings often show dark clouds with birds, such as pied crested cuckoos from east Africa, flying in front as the traditional sign that these clouds are bringing the rains.

▶ For some Asian peoples, a five-toed dragon is the bringer of storms and rain.

▶ During July the Sun lies north of the equator. India and the rest of the Asian continent heat up in the long days of strong sunshine. Warm air rises above the land and pulls in moist air from the cooler ocean like a giant sea breeze. These incoming winds bring heavy rains to the land. In January the Sun has moved south of the equator. Australia warms in its summer sunshine while Asia cools during its winter. Dry, chilly winds now blow outward from Asia. They gather warmth and moisture from the oceans and bring summer rains to northern Australia.

wet south–westerly winds

Summer rains

For many farmers in Asia the monsoon simply means their rainy season—lasting from June to September each year. While Americans and Europeans divide the year into four seasons, the people of south Asia think only of two (the dry and wet seasons) or sometimes three (the cold, hot, and rainy seasons).

People in India suffer several months of very hot, dry weather before the monsoon rains arrive. Wells empty, reservoirs dry up, and the soil becomes baked hard until it cracks in the dry heat. Everyone longs for the start of the rains.

The southern tip of India is the first place to feel the rains and all of India waits to hear officially that the rains have begun. There are many false starts. Dark clouds in the distance may turn out to be a dust storm or a swarm of locusts. One or two brief thunderstorms may bring false hope. But around June 1 the monsoon rains finally arrive and move steadily north until they have reached all of India by mid-July.

▲ **This dark sky over Nepal heralds** the coming of the summer monsoon. The monsoon occurs in parts of Asia, Australia, and Africa, but Asia is so vast that its monsoon is the strongest. It is the most important, too, because it affects the lives of so many people.

The rains are greeted with great joy and excitement. Children sing traditional songs sitting in the rain on swings that are hung from trees and decorated with flowers.

The children's songs are about the rains having returned and the promise of good fortune that this means. Bare trees turn green and grass begins to grow. The rains bring hope for the people. At last the plowing can start and the rice can be planted. If the rains are good, there will be no famine that year.

But the monsoon brings worries, too. Will there be enough rain for that year's crops? Will too much rain fall causing floods that wash away homes and crops? The monsoon affects everyone's lives.

WINTER MONSOON

Himalayas

cold land mass creates high pressure

no rain

dry northeasterly winds

heavy rain

clearing skies as rains move inland

INDIAN OCEAN

INDIAN OCEAN

INDIAN OCEAN

NORTH

wet south-westerly winds

Flash floods

Moving as fast as an express train, there is little warning before a flash flood arrives. It is usually the result of heavy rainfall from a thunderstorm or hurricane over mountains, but a very rapid snowmelt or a dam failure can create flash floods too. The water level in a mountain stream rises very quickly and overflows its banks, sending a towering wall of water sweeping down the valley. Swift action is needed to save lives. Seconds count. There is often little to do but flee for high ground.

Only a small area is usually affected, but the results are devastating. When 12 inches of rain fell over four hours in Big Thompson Canyon in the Colorado Rockies in July 1976, a 19-foot-high flood wave destroyed 418 houses and killed 139 people. Even worse was the flash flood of June 1972 in the South Dakota Black Hills, which killed 236 people.

A wall of water

In August 1996, a flash flood tore through a mountain campsite in the Pyrenees in Spain after thunderstorms had produced 3 inches of rain in only two hours. The rain charged down the steep mountain slopes gathering soil, stones, and boulders, and uprooting trees. The mountain stream became a raging torrent. Bridges caught some of the debris and blocked the stream so that the water level rose even higher behind them.

bridge

narrow valley

▶ **A flash flood**
is triggered when a large amount of rain falls in a very short time over a mountainous area. Little or no water seeps into the soil as it rushes down the steep slopes. So the water level in a mountain stream rises very quickly, sending a flood wave roaring down the valley battering anything in its path.

◀ **A campsite at Virgen de las Nieves**
in the Spanish Pyrenees was devastated in August 1996 as caravans, tents, and cars were swept away by a flash flood. Some people survived by hanging onto trees.

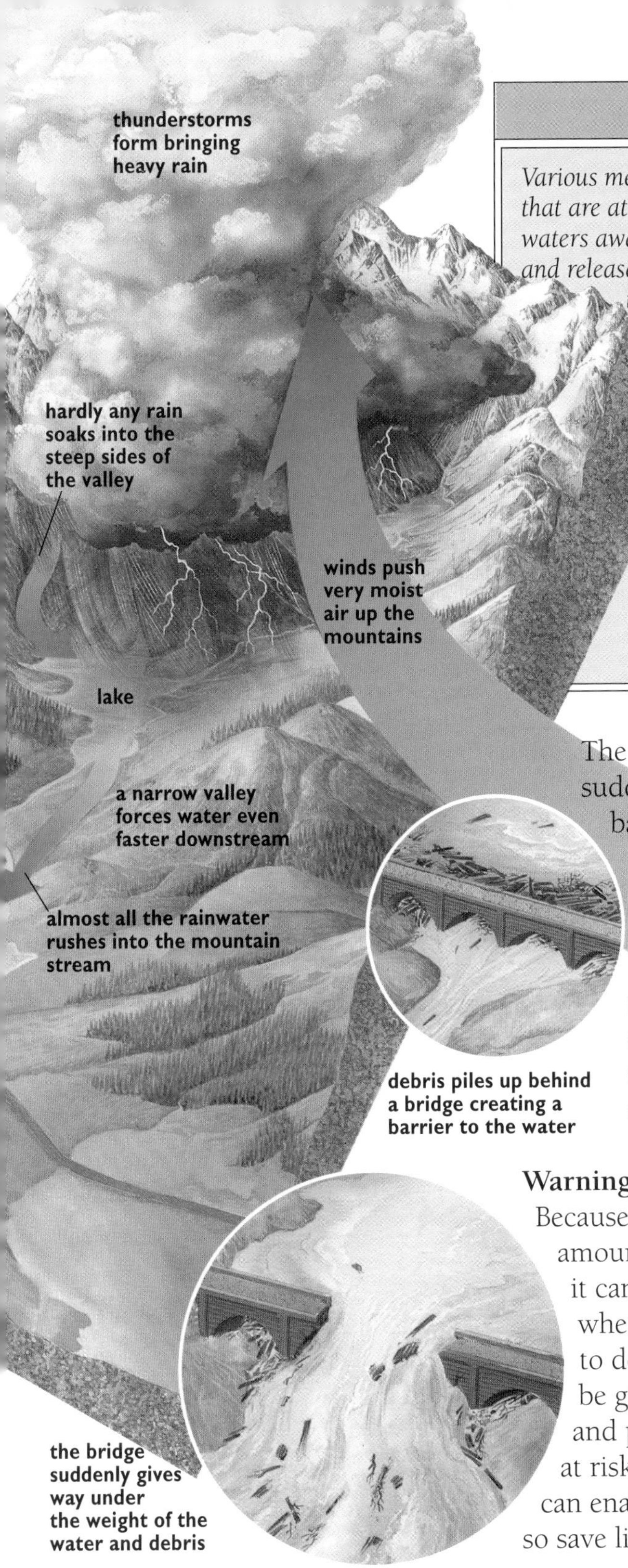

Labels on main illustration (left/top)

thunderstorms form bringing heavy rain

hardly any rain soaks into the steep sides of the valley

winds push very moist air up the mountains

lake

a narrow valley forces water even faster downstream

almost all the rainwater rushes into the mountain stream

debris piles up behind a bridge creating a barrier to the water

the bridge suddenly gives way under the weight of the water and debris

PROTECTION AGAINST FLOODS

Various methods are used to help prevent floods in areas that are at risk. New channels can be added to carry flood waters away from towns. Dams hold back flood waters and release them slowly. Trees planted on steep slopes help to slow the water's downhill passage, allowing more to seep into the soil.

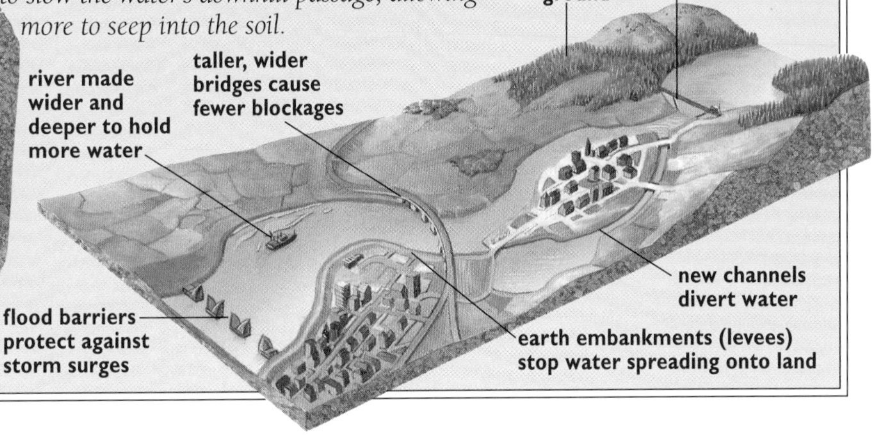

trees encourage water to seep into ground

dams control water levels

river made wider and deeper to hold more water

taller, wider bridges cause fewer blockages

flood barriers protect against storm surges

new channels divert water

earth embankments (levees) stop water spreading onto land

The increasing weight of water suddenly broke through the barriers sending a giant wall of water crashing down the valley. At least 62 people died and 180 were injured when the flood wave smashed into caravans and cars. Some people were swept more than 10 miles down the valley.

Warnings

Because radar can measure the amount of rain as it is falling, it can be used to determine whether a flash flood is likely to develop. Warnings can then be given via the radio, sirens, and police megaphones to people at risk. Even 20 minutes' warning can enable people to evacuate and so save lives and belongings.

River floods

Unlike flash floods, river floods are slow to develop. They are often seasonal and are caused by long periods of heavy rainfall or the melting of deep snow over large river basins. Warnings can be given many hours or days before a river threatens to overflow its banks.

River floods can affect a vast area of land. The Mississippi flood in the summer of 1993 hit nine states. Crops were lost, roads were washed away, and buildings were flooded and badly damaged. Nearly 70,000 people were left homeless; 50 people lost their lives. The damage cost an estimated $12 billion.

Unlike flash flood waters, which subside very quickly, a river flood can remain for days or even weeks. The Mississippi flood lasted from June to August.

Deep freezes

When heavy snow is driven by gale force winds in temperatures below the freezing point, it is called a blizzard. Traveling in a car in a severe blizzard is impossible, even if the engine keeps running, because it is difficult to see past the hood.

A blizzard and the bitterly cold weather and deep snowdrifts that follow can bring chaos to an entire region. Airplanes are grounded and cars, trucks, and trains grind to a halt for days or even weeks. People may be trapped in their homes, and electricity supplies can be cut off by power lines being toppled by the heavy, gusting snow or by branches snapped off by the weight of the snow.

In 1994, winter temperatures in the northeastern United States were the lowest recorded for over 100 years. Temperatures in parts of Minnesota fell to −40°F. Around 100 people died from the cold, from heart attacks as they shoveled away the snow, and from traffic accidents caused by the bad weather.

▲ **On steep slopes, fresh, powdery snow** may break loose from a firm layer of older snow beneath it. The avalanche tumbles down the mountainside at speeds of 200 mph, smashing aside trees and can bury a village in seconds.

▶ **Snow makes the landscape look** beautiful, but when the snow is very deep, it is a dangerous form of beauty. People become trapped in their homes and cars, the low temperatures bring a deadly wind chill, and power is disrupted.

Deep snows

Storms need a lot of moisture to produce deep snow. When strong arctic winds sweep across the Great Lakes they pick up so much warmth and moisture that they can dump 8 feet of snow on cities downwind.

The city of Buffalo lies at the end of Lake Erie. During a five-day snow storm in January 1977, with winds gusting to 68 mph, some parts of the city and surrounding area were buried under snow drifts 30 feet high. More than 5,000 cars and trucks were abandoned on major highways. In December 1995, the lake produced a similar effect and Buffalo received 38 inches of snow in 24 hours. That is a lot of snow, but it is only about half the world record over 24 hours, which was suffered by the village of Montague in New York State in January 1997.

Avalanches

An avalanche can be triggered by heavy snowfall, a slight thaw in the snow, or a strong gust of wind. Even the noise or movement made by a skier can set off an avalanche—in just the same way as someone banging on a front door can get covered by partly-melted snow falling off the roof above.

There are two kinds of avalanche: the loose snow type and the slab type. Loose snow behaves like dry sand. If the snow is compact (with the grains of snow tightly packed together), it falls in a great slab.

AVALANCHE PROTECTION

Fences, walls, and rakes are built to hold the snow on the mountainside and help prevent an avalanche. But, once an avalanche starts, little can be done to stop the snow surging down the mountainside. Trees and mounds can help to slow the avalanche, and walls and wedge-shaped constructions can be built to divert the snow around buildings. Some buildings are even designed in the shape of a ship's bow to divert the snow to either side.

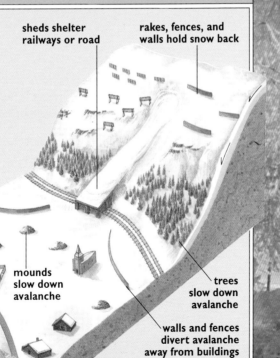

sheds shelter railways or road

rakes, fences, and walls hold snow back

mounds slow down avalanche

trees slow down avalanche

splitting wedges and wedge-shaped buildings divert avalanche around sides

walls and fences divert avalanche away from buildings

When the rains fail

Farming is geared to the amount of rain that usually falls each year. When rainfall is much less than expected for the time of year, a period of unusual dryness known as a drought sets in. Drought can happen in areas that normally receive a lot of rainfall as well as in those areas that receive very little. If only one-quarter of the year's normal rainfall fell in south Florida, the region could be said to be suffering a drought. Yet, if the same amount—15 inches—fell in the Great Plains, this would be considered quite normal. Long periods of unusual dryness can cause huge problems.

▲ **During periods of drought,** Australia seems to suffer worse wildfires than the United States partly because the oils of Australia's common eucalyptus tree help to fan the flames. Pieces of burning eucalyptus bark may be carried by winds for several miles, starting many new, deadly fires.

Drought

The United States grows much of the world's grains. When drought occurs here, the rest of the world is affected. The 1988 drought was one of the worst in 50 years. A high pressure area known as a blocking high (*see right*) hung over the U.S. and diverted away the storms that normally bring the rains. The drought affected 30 states with the Midwest and the South suffering most. Crops such as barley, wheat, corn, and soybean suffered greatly.

By mid-1988 many areas had received less than one-quarter of the normal rainfall expected for the first six months of the year. Temperatures rose to 104°F in the shade, and the water level in the Mississippi fell so low that hundreds of barges were grounded. Forest fires burned and spread out of control in seven western states.

The 1988 drought was the worst in the United States since the 1930s when six years of drought and dust storms ruined the landscape of the Midwest. In some places, not a blade of grass could be seen. Great clouds of blowing topsoil often blotted out the Sun. The drought forced many

jet stream

trough of low pressure

ridge of high pressure

low pressure area

blocking high

low pressure area

jet stream snakes far north and eventually splits into two—the southern stream becomes stronger, the northern stream gradually dies out

▲ **The frontal depressions** or lows that bring rain to the middle latitudes are steered by a jet stream (*see pages 18–19*). If the jet curves around in a giant loop, the lows follow, too. A large anticyclone called a blocking high forms inside the loop. This brings clear skies and hot, dry weather in summer. If it stays for many weeks, it can bring drought.

farmers in Kansas, Oklahoma, and Texas to flee parched, dusty fields for the green promise of California.

Europe suffers drought, too. The period from May 1975 to August 1976 was one of the worst droughts in hundreds of years. The area affected stretched from Scandinavia to western France. Southern England recorded only half its normal precipitation for this period. Reservoirs dried up completely and crops failed.

In contrast, Iceland and the northern Mediterranean region received one and a half times their normal rainfall because the jet stream steered the rainstorms in their direction.

Australia suffers frequent droughts. Officials step in quickly to help farmers. Water and animal feed are sent in by trucks or trains, deeper wells are sunk, and tens of thousands of sheep and cattle may be transported to wetter areas.

▼ Reservoirs can dry up completely during a drought and emergency supplies of drinking water have to be found. Crops wither and can fail, leading to severe food shortages. The parched and dusty soil is stirred up into huge dust storms, and fires feed on the dried out vegetation.

LONG-TERM DROUGHT

Deserts such as the Atacama Desert in northern Chile have not seen rain for tens of years, but this is normal. Along the southern edge of the Sahara Desert in northern Africa, however, a region called the Sahel has suffered drought on and off since the late 1960s. This is not normal, and famine has brought hunger and misery to millions of people.

Droughts can be caused by people interfering with the water cycle. Since the 1960s, water has been taken from the rivers flowing into the Aral Sea in central Asia to irrigate expanding areas of cotton and rice. Thirty years later, the Aral Sea, once the fourth biggest lake in the world, has shrunk to half its original size. Summers in the region are much drier and hotter than ever before. The number of days without rain has increased five-fold and the region suffers frequent dust storms.

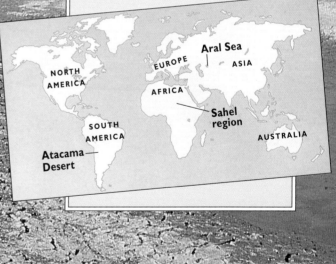

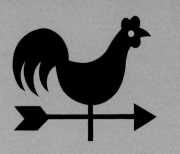

Collecting information
44–45

Making weather forecasts
46–47

Using weather forecasts
48–49

Skywatching
50–51

Cloudscapes
52–53

WEATHERWISE

BEING ABLE TO FORECAST ACCURATELY HOW THE WEATHER WILL CHANGE is the big challenge for meteorologists, the scientists who study the weather. Millions of measurements of weather conditions across the world are the first steps toward producing weather maps for the days ahead.

Collecting information

Around 12,000 weather stations across the world send weather observations every three hours to a number of international weather centers. Measurements taken by specially-equipped airplanes, ships, balloons, and buoys that drift in seas and oceans fill in missing gaps, while satellites and radar provide pictures of the broad pattern of clouds and storms. Knowing the current weather even in the remotest parts of the world is vital, since what happens in one place often affects the weather in another.

▲ **Specially equipped airplanes** collect and send back weather information from various levels in the atmosphere. Certain airplanes can fly into the heart of violent storms.

Satellite pictures

High above the Earth's surface, satellites continually photograph clouds and storms and measure moisture and temperature at various heights in the atmosphere. There are two main types of weather satellite. Five geostationary satellites orbit the Earth 22,500 miles above the equator. They travel at the same speed at which the Earth rotates and so always remain above the same spot on the Earth's surface. They send back pictures every 30 minutes.

Polar orbiting satellites travel from pole to pole at a height of only 530 miles. They produce very detailed pictures of weather systems and pass over the same place on Earth twice a day.

There are two types of satellite picture. One is a normal photograph, which can be taken only in daylight. The other is an infrared picture which can be taken day or night. It measures the temperature of the cloud top or, where the sky is clear, the Earth's surface. Warm land shows as a black or dark gray area and cold cloud tops as light gray or white. Since air temperature decreases with height in the weather layer, the higher the cloud top, the whiter it appears in the picture.

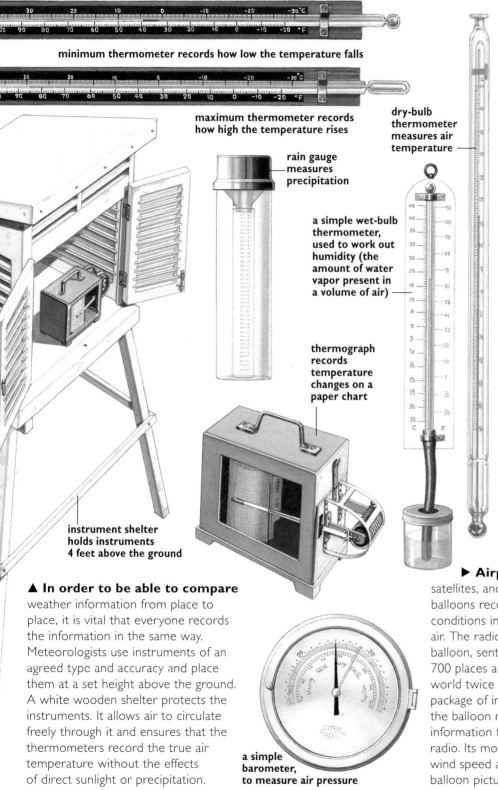

minimum thermometer records how low the temperature falls

maximum thermometer records how high the temperature rises

rain gauge measures precipitation

dry-bulb thermometer measures air temperature

a simple wet-bulb thermometer, used to work out humidity (the amount of water vapor present in a volume of air)

thermograph records temperature changes on a paper chart

instrument shelter holds instruments 4 feet above the ground

Meteorologists record information about each element of the weather (see below). The type of precipitation is noted, but the amount is expressed as the depth of water collected in a rain gauge after any snow, sleet, or hail has melted. A snow depth of 12 inches melts down to around 1-inch depth of water.

clouds

temperature

sunshine

humidity

wind strength and direction

air pressure

visibility

precipitation (rain or snow)

▲ **In order to be able to compare** weather information from place to place, it is vital that everyone records the information in the same way. Meteorologists use instruments of an agreed type and accuracy and place them at a set height above the ground. A white wooden shelter protects the instruments. It allows air to circulate freely through it and ensures that the thermometers record the true air temperature without the effects of direct sunlight or precipitation.

a simple barometer, to measure air pressure

▶ **Airplanes,** radar, satellites, and radiosonde balloons record weather conditions in the upper air. The radiosonde balloon, sent up from 700 places around the world twice daily, carries a package of instruments. As the balloon rises, it sends back information from various heights by radio. Its movement is tracked to record wind speed and direction. The type of balloon pictured here is called a Jimsphere.

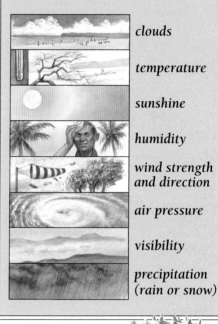

Making weather forecasts

Just by looking out of the window, you can forecast what the weather will be like in 15 minutes' time. If there is no cloud in sight, a confident prediction would be that it will not rain. Forecasting whether it will rain in six hours' time needs information from a satellite or radar showing whether a storm is on its way.

Forecasts for the following day or for several days ahead rely on information about storms that are currently many thousands of miles away—how will they move and change, and will any new storms develop? This requires detailed information about air temperature, moisture, pressure, and winds and an understanding of how they are likely to affect each other. Sorting through such information becomes very complicated and meteorologists need powerful computers to help them produce weather forecasts.

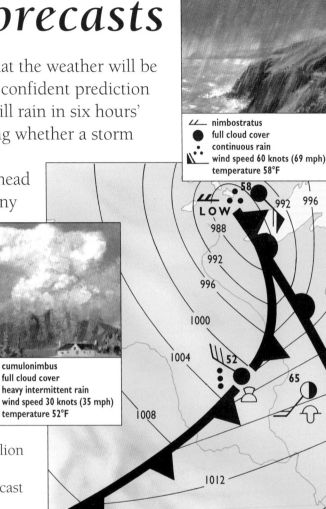

nimbostratus
full cloud cover
continuous rain
wind speed 60 knots (69 mph)
temperature 58°F

cumulonimbus
full cloud cover
heavy intermittent rain
wind speed 30 knots (35 mph)
temperature 52°F

How forecasts are made

A weather forecast for the next day is made up from information collected from across the world, which is fed into a supercomputer. The computer works out how air pressure, winds, temperature, and moisture are likely to change at lots of places at different heights in the atmosphere but for only 10 minutes ahead. The weather predictions produced by this very short forecast are then used by the computer to forecast another 10 minutes ahead. This process is repeated until a forecast for 24 hours ahead is produced. Supercomputers can work out several billion sums a second and can complete a weather forecast in only 20 minutes.

Forecasting weather accurately two or three weeks ahead is almost impossible. This is because forecasters cannot consider all the local weather variations and the resulting changes that can occur everywhere. For example, it may be almost by chance that a small cumulus cloud grows into a thunderstorm, which develops into a raging hurricane, which then affects weather developments elsewhere in the world.

stratocumulus
half cloud cover
wind speed 15 knots (18 mph)
temperature 65°F

Reading weather maps

The continuous lines on weather maps are the isobars, which join up areas with the same air pressure (*see also page 14*). Air pressure is measured in millibars—the average pressure on the Earth's surface is 1013 millibars. Like an ordinary map, which uses height contours to show hills and valleys, a weather map uses isobars to show high and low pressure areas across the Earth's surface. If the pressure in a low decreases from day to day, it means that the low is strengthening and the stormy weather will worsen.

Winds blow clockwise around a high and counterclockwise around a low in the northern hemisphere (they blow in the opposite direction in the southern hemisphere). Generally, winds blow along the line of the isobars, but they turn slightly toward a low and away from a high. The closer together the isobars, the stronger the winds will be.

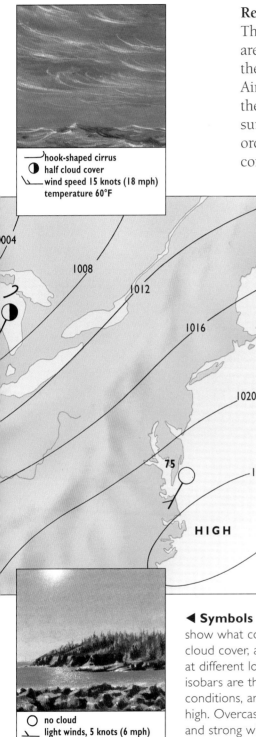

hook-shaped cirrus
half cloud cover
wind speed 15 knots (18 mph)
temperature 60°F

1004
1008
1012
1016
1020
75
1024

HIGH

no cloud
light winds, 5 knots (6 mph)
temperature 75°F

◀ **Symbols used on weather maps**
show what conditions—such as wind speed, cloud cover, and precipitation—can be expected at different locations. The areas enclosed by isobars are the highs and lows. Clear skies, dry conditions, and very light winds will be felt in a high. Overcast, cloudy conditions, precipitation, and strong winds occur in lows. The fronts within a low produce a particular sequence of weather changes (*see pages 18–19*).

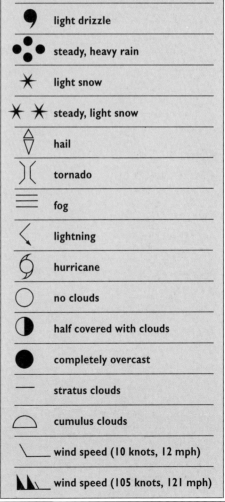

WEATHER SYMBOLS

Meteorologists around the world use the same set of weather symbols so that they can exchange information easily. The meaning of some symbols is not obvious to the public, so the symbols used on television and newspaper weather maps tend to be simpler.

Meteorologists measure wind speed in knots: 1 knot is 1.152 mph. The wind symbol is drawn pointing toward the cloud cover symbol in the direction in which the wind is blowing.

SOME INTERNATIONAL WEATHER SYMBOLS:

	light drizzle
	steady, heavy rain
	light snow
	steady, light snow
	hail
	tornado
	fog
	lightning
	hurricane
	no clouds
	half covered with clouds
	completely overcast
	stratus clouds
	cumulus clouds
	wind speed (10 knots, 12 mph)
	wind speed (105 knots, 121 mph)

Using weather forecasts

Knowing how the weather will change later today, tomorrow, and next week can help us plan our activities. Weather forecasts help us decide what clothes to wear and whether it will be better to play or work outdoors or indoors. People whose work can be severely affected by the weather, such as farmers and builders, can lose money if they do not plan their work according to what the weather will be like. The lives of people traveling in cars, trains, airplanes, and ships can be made safer if warnings are given of severe weather such as fog, gales, icy conditions, or heavy rains.

▲ **Oil rigs may be damaged** by violent storms at sea. Forecasters advise of dangerous weather conditions, enabling workers to be airlifted to safety. For ships, rough seas mean longer journeys, which use up costly fuel. Captains rely on forecasts to plan routes to keep their ships away from danger.

highway officials need to know when to sand and salt roads

fishing boats and cargo ships use routes that avoid strong winds and storms

The benefits of forecasts

Many industries are affected by the weather and can benefit from forecasts. What we buy and when we buy it may be influenced by the weather. Supermarkets and shops use forecasts to work out likely sales patterns and adjust their stocks and displays accordingly. This includes sales of clothing, food and drinks, and paint for house exteriors.

When the temperature falls, the demand for electricity and gas for home heating rises. Power companies use temperature forecasts to plan how much power will be needed and to ensure that electricity-generating plants or gas supplies can meet that demand.

On construction sites, forecasts help workers to avoid problems that bad weather can bring. Fresh concrete can be damaged by frost and rain. Cranes are too dangerous to operate and scaffolding is unsafe to erect if strong winds are expected.

supermarkets order food stocks and arrange displays according to the weather

motorists need warnings of fog, drifting snow, icy roads, strong winds, and heavy rain since these can cause accidents

farmers need to know the best time to sow seeds, spray fertilizers, and water and harvest crops

warnings of strong winds, poor visibility, and storms allow hikers and climbers time to change their route or seek shelter

tall cranes on building sites cannot be used safely in strong winds

▲ Weather forecasts, providing they are accurate, can help make our lives more comfortable, safer, and less costly. Most people use the general forecasts that are available from television, radio, and newspapers. Others, especially those whose work is heavily influenced by the weather, pay weather agencies to supply the details they need.

forecasts enable airplanes to avoid strong headwinds and turbulence

Severe weather warnings

The effects of severe weather can be very costly, both in terms of human lives and damage to property. Dense fog, heavy snow or rain, icy conditions, and strong winds make roads dangerous and lead to accidents. Large trucks can be overturned by strong, gusty winds so highway officials need to know when to close high bridges to such vehicles.

Icy roads are made safer when spread with sand or salt. Sand helps tires grip the road surface, while salt lowers the freezing point of water and prevents ice from forming or makes it melt. Salt works best if spread before ice has formed—this requires an accurate forecast.

Airplanes may be at risk when landing in fog, low clouds, downbursts, or strong winds—especially winds blowing across a runway. Warnings allow the pilot to delay the landing or divert the airplane to another airport.

Early warnings of gales or storms give fishing fleets time to seek shelter. In the colder oceans, warnings of strong winds and freezing temperatures are crucial. In these conditions, sea spray can form thick, icy layers on deck structures which could cause a boat to capsize.

WRONG FORECASTS

Sometimes weather forecasters get it wrong. Newspaper headlines read "Why weren't we warned?" when winds gusting to 120 mph hit southern Britain in October 1987 (see right). The storm caused 26 deaths and brought down 15 million trees. Such big mistakes happen rarely. They can occur when the regional, national, and global computer-generated forecasts that are used by meteorologists do not agree with one another. The senior weather forecaster must then decide, based on experience, which forecast is more likely to be right. Sometimes the wrong choice is made.

Skywatching

For thousands of years, people have looked to the sky for clues about how the weather may change. They have observed how certain types of weather follow one another and have watched the changes that occur after rainbows, halos, and red skies. Our ancestors made up sayings, which often rhymed, to help them remember the more useful weather signs. Even today, there is much to learn from skywatching. *But remember not to look directly at the Sun, even with sunglasses on.*

Blue skies

The white light from the Sun is made up of all the colors of the rainbow mixed together. As sunlight travels through the atmosphere, it strikes water droplets, ice crystals, or solid particles of dust and pollution. These make the light and the colors within it behave in particular ways, which can tell us what is happening in the atmosphere.

The molecules of gas that make up air scatter sunlight in all directions. The blue in sunlight is scattered the most, creating a brilliant blue sky on a sunny day. But if there are tiny particles of pollution or dust in moist air, these particles absorb moisture and swell to a size that scatters all the colors of light equally. On a clear day this makes the sky turn milky white.

The sky shows its deepest blue when the air is cold, dry, and clean, with no pollutants present.

Lights in the sky

The sky can display unusual and delicate, white or colored shapes of light. These are the effects of sunlight striking water droplets or ice crystals.

As light passes through a droplet or crystal, it may be bent (or refracted). Each color in the light is bent at a slightly different angle, so the colors become separated. This effect creates one of the most beautiful spectacles in the sky: the rainbow (*see above right*). A rainbow forms when sunlight from behind you is split into its colors by raindrops falling in front of you. Because weather usually moves from west

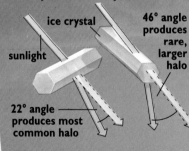

sunlight — ice crystal — 46° angle produces rare, larger halo — 22° angle produces most common halo

◄ **Rings of light around the Sun,** called halos, are created when light is bent by column-shaped ice crystals in the air. The size of the halo depends on the size of the angle at which the light is bent.

A rainbow forms when sunlight is refracted on entering a water droplet, then reflected inside and finally refracted again on leaving. A secondary rainbow may form outside the main, or primary, one when sunlight reflects twice inside raindrops. The colors are brightest and clearest at the feet of a rainbow, where the largest raindrops occur.

sunlight

light reflected

light refracted

water droplet

light separated into colors of rainbow

to east in the middle latitudes, a rainbow seen to the east in the afternoon or evening may indicate that good weather is on its way. The presence of sunshine means that clear skies lie to the west, while the rain in the east has already passed. A morning rainbow in the west indicates that rain may arrive soon.

As sunlight strikes ice crystals it may be bent to form bright spots or mock Suns, called sun dogs, on either side of the Sun. Sometimes light bounces (or reflects) off ice crystals to form an orange or red sun pillar. This is a column of light above a setting or rising Sun. Because cirrus clouds contain ice crystals, the appearance of sun dogs, sun pillars, or halos (*see far left*) may be the first sign of an approaching warm front.

RED SKIES

The sky at sunset and sunrise often turns a brilliant orange-red. This is because the sun is low in the sky and its light has to travel a longer distance through the atmosphere. The sunlight is split into its colors and scattered (see page 50) as it makes its journey. But because the blue light is scattered most, only red, orange, and *yellow are left by the time the sunlight reaches the ground.*

A red sky in the evening is often thought to be a sign that fine weather will come next morning. This is because a red sunset indicates that the sky to the west is clear and there are no clouds on their way bringing rain.

Sun

atmosphere

at midday, light travels through less atmosphere

at sunrise and sunset, light travels farther through atmosphere

Sun

Cloudscapes

▼ **Waves may develop in winds as** they blow over a mountain range. Smooth, lens-shaped clouds, called altocumulus lenticularis, form at the tops of the waves where the air rises, cools, and condenses into water droplets.

wind —— lenticular clouds

wind

circular flow of air makes cloud deeper close to mountain

mountain range

▼ **Lenticular clouds usually remain** in the same spot for a long time. When there are dry and moist layers in the winds, the clouds may pile up like a stack of plates.

The many different shapes, sizes, textures, and colors of clouds, as well as how they change and what patterns they form, can provide further clues about how the weather might change. Simply knowing that clouds are blown along by the wind tells us the direction from which the weather is coming. The speed of the clouds tells us how quickly weather changes may occur.

Some clouds are signs of what is happening on a local scale. A line of cumulus clouds developing along the leading edge of a sea breeze marks the movement inland of cool, moist sea air. Such clouds give clues about weather changes over several hours. Other clouds are the result of what is happening in the atmosphere across thousands of square miles over several days. These may be part of an enormous frontal depression or hurricane.

Will it rain?

Clues about whether it will start or stop raining soon can be found in the sky and the clouds. A few hours before rain arrives on a warm front, the air is often unusually clear. Distant objects stand out distinctly and look brightly colored. This is because cold, dry polar air often moves ahead of a warm front and gives very good visibility. There are many local weather sayings that refer to being able to see a particular landmark very clearly, such as a distant hill, when rain is on its way.

In the middle latitudes, where weather fronts bring bands of rain, people often say, "Rain before seven,

CONTRAILS

The long white threads of cirrus clouds streaming behind aircraft high in the sky are called condensation trails or simply contrails. They are formed when water vapor from the jet exhausts turns into ice crystals in the freezing air. Contrails can be a weather clue. Only short contrails are seen when the air is dry because they evaporate quickly. Lengthy and long-lasting contrails form when air is moist and this is a sign that a frontal depression is on its way.

fine before eleven." Many rain bands are 160–200 miles wide and move at about 40 mph, so we can expect it to rain for four or five hours.

Deep clouds are more likely to produce rain than shallow clouds because they contain more water. You can tell the depth of a cloud even when you cannot see its top: its base will appear gray or black if the cloud is deep because the water blocks some of the sunlight.

Some clouds in the distance may show a dark fringe hanging from beneath. The fringe is called virga and it is rain or snow falling but evaporating in very dry air before it reaches the ground. Showers may soon follow because the virga may eventually moisten the underlying dry layer sufficiently for showers to reach the ground.

Sunny days often begin clear but by late morning small fluffy white cumulus clouds may appear. These are formed by warm air rising from

sun-warmed patches of ground (*see also pages 10–11*) and are called fair weather cumulus. Occasionally they produce rain if the rising air is moist and warm enough. If cauliflower-shaped bulges start growing rapidly from the cloud tops, this is a sign that showers may follow in the afternoon. Usually, however, the clouds simply shrink and eventually disappear as the day cools with the setting sun.

▶ **The underside of a** thunderstorm's anvil may display large, round bulges, called mamma (which is the Latin word for "breast"). They occur in the front or rear anvil when pockets of the cloud fall slowly but are supported by warm air rising underneath.

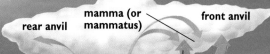

rear anvil   mamma (or mammatus)   front anvil

upcurrent

warm, rising air pushes against cool, sinking air forming bulges

Climate clues
54–55

Ice ages
56–57

Darkness and flickering light
58–59

El Niño
60–61

Polluting the atmosphere
62–63

The greenhouse effect
64–65

THE WORLD'S CHANGING CLIMATE

THE EARTH'S CLIMATE IS FOREVER CHANGING. SOMETIMES THE CHANGE IS slight, perhaps with dry summers or mild winters becoming a little more frequent. Other changes are more dramatic, such as when the planet plunges into an ice age.

Climate clues

Detailed records of climate exist only for the past 150 years or so, since it was not until the 19th century that accurate weather information began to be collected from around the world. To learn about climates thousands or millions of years ago, climate clues must be found.

Animals and plants adapt to the climate in which they live. If the remains, or fossils, of animals and plants living in the past can be found, dated, and identified as having lived in warm or cold conditions, then a picture of climate can be created.

Layers of evidence

Some evidence provides only a brief glimpse of what climate was like at one place at one time. What is needed, however, is a long and complete record of how climate has changed.

A tree adds a new ring of wood to its trunk each growing season. The narrower the ring, the drier and colder that year has been. Using living trees and the remains of old tree trunks, scientists have constructed a year-by-year climate record going back some 8,000 years.

Similar time slices can be found in samples of rock drilled from the seabed. These contain the shells of different types of tiny sea creatures that lived long ago. Some liked warm conditions while others preferred cold, so the rocks show climate changes. Pollen and insects preserved in lake mud, and animal and plant fossils exposed in rocks along coastal cliffs provide further clues.

▲ **Flies and spiders trapped** in the sticky sap of fir trees can be preserved for millions of years when the sap hardens into amber. Air bubbles caught in the amber can tell us how the amount of each gas in the atmosphere has changed.

▼ **Fossils show us what animals and plants** from long ago were like. When a fish dies, it sinks to the seabed and decays, leaving only the hard bones and teeth (1). Sand settles around the bones. Over millions of years the chemicals of the bones and teeth dissolve, but are replaced by rock which takes on the same shapes, and forms the fossil (2). Movements of the land and the slow wearing away of the surface gradually exposes the fossil (3).

3

2

▼ **The thin annual growth rings** in a tree show cold years and drought years and can be dated accurately.

HISTORICAL EVIDENCE

Clues about climate come from many sources. People have kept weather diaries, ships' log books, and records of the dates of wine harvests and first frosts. All of these reveal important information about past climates. Old photographs and paintings give us pictures of weather conditions, such as the deep snows of northern Europe in this 16th-century painting. Cave paintings by early peoples show that cold-loving animals once roamed lands that now enjoy a warm climate.

Causes of climate change

The brightness of the Earth's surface has a major effect on climate. The brighter and shinier the planet, the more sunlight will bounce (or be reflected) off its surface and the cooler the Earth will be. Dust veils produced by volcanic eruptions, smoke from cities, and expanding deserts make the planet shinier and therefore cooler.

The movement of continents and the creation of mountains over millions of years change the position and strength of ocean currents. These affect the transfer of warmth around the planet and so change the global pattern of winds and storms.

Dinosaur extinction

If a large enough asteroid (a rocky body from space) struck the Earth it would throw billions of tons of vaporized rock into the atmosphere, blotting out the Sun and causing the global temperature to plummet. A freezing winter lasting several thousand years would follow. Such a strike may have caused the sudden extinction of the dinosaurs and other animals some 65 million years ago, when an asteroid measuring 10 miles wide struck the Gulf of Mexico. It left a crater 180 miles across.

▲ **Bubbles of air, dust, and chemicals** trapped in ice cores cut from Antarctic and Greenland ice sheets give clues about the climate tens of thousands of years ago.

Ice ages

During the past two million years, the Earth's climate has plunged in and out of icy conditions as many as 20 times. The global climate has swung from the warm conditions we enjoy now, called interglacial, to bitterly cold conditions, called glacial, and back again. The last glacial period ended about 10,000 years ago. At its height—about 18,000 years ago—Canada, northern parts of the United States, northern Europe, parts of Siberia, New Zealand, Tasmania, and the tip of South America were covered by thick ice sheets, some 3,000 feet deep. Today, large ice sheets are found only in Greenland and Antarctica, but the ice of glacial periods has left its mark. It carved new valleys and lakes and pushed sand and stones into new shapes on the surface of the land.

▲ The axis on which the Earth spins is now tilted at 23.5° but every 41,000 years or so the angle changes by a few degrees.

▶ The Earth's orbit stretches from a circle to an oval over a period of 100,000 years. Its axis changes direction and traces out a circle every 21,000 years, so the planet behaves like a wobbling spinning top.

A glacial begins

The almost regular occurrence of glacials is caused by changes in the Earth's orbit and the axis on which the planet spins (*see left and below*). These changes alter the amount of sunlight that reaches different parts of the Earth so that about every 100,000 years the climate begins to cool. Winters become longer and more snow falls, while shorter summers mean that less snow melts. During hundreds to thousands of years, the snow builds up to form thick ice sheets high up in mountains. The ice spills over to cover entire countries, sending tongues of ice called glaciers down valleys and dumping icebergs into the oceans when they reach the coast. Sea ice thickens, and the planet becomes shinier (*see page 55*), reflecting more sunlight and cooling the planet

movement of axis

Sun

N

Earth

circular orbit

oval (elliptical) orbit

▲ **Only Greenland and Antarctica** have large ice sheets today, but some mountains have small glaciers like this one in the Alps in Switzerland.

northern ice sheets and sea ice at height of last glacial

northern ice sheets and sea ice today

even further. Only changes in the Earth's orbit and axis of rotation can drag it out of its deep freeze.

Ice Age climate

At the height of the last glacial, the Earth's climate was very different from today. The northern continents were covered in dome-shaped ice sheets several thousand feet thick— the depth of the ice that covers Greenland today. Blustery cold winds swept down these icy slopes chilling the surrounding land and freezing the soil.

The sea level fell by 330 feet as water became locked up in growing ice sheets. Coastlines extended into

◄ **About 18,000 years ago,** the ice sheets in the northern hemisphere had spread across the northern half of North America and much of northwest Europe. In the southern hemisphere, only Antarctica, New Zealand, Tasmania, and the tip of South America experienced much expansion of ice sheets.

the sea by 150 miles or more in places. Islands like Britain became joined to Europe by land, while Alaska became linked to Siberia.

Summers were brief. Some of the soil thawed and grass grew. Edges of the ice sheet melted and torrents of meltwater spread out from under the ice. The water carried sands, gravel, stones, and boulders that the advancing ice had bulldozed from the land. This material was then dumped, forming distinctive shapes such as snakelike ridges called eskers.

BODIES IN ICE

During glacials, elephant-like creatures called woolly mammoths roamed near the ice sheets. The mammoth's long shaggy coat and layer of blubber 3 inches thick protected it from the cold. Some had curled tusks 14 feet long. The mammoth calf shown below was found preserved in the frozen soil in Siberia in northern Asia. It is rare to find preserved bodies, but even human bodies from thousands of years ago have been discovered.

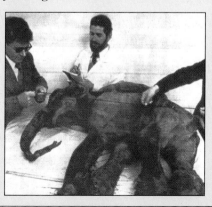

Darkness and flickering light

When a volcano erupts, it throws huge amounts of ash, dust, and gases high into the atmosphere. The material spreads out and acts like a sunscreen, stopping some of the sunlight from reaching the ground. Chilly winters and cool summers may follow a major volcanic eruption.

Volcanic sunscreens are not the only reason for less sunlight reaching the Earth's surface. The amount of light the Sun gives out can change. When the Sun's light is weak, the Earth may cool.

Volcanic eruptions and climate

The heavy ash from a volcanic eruption is normally washed out of the troposphere within a few weeks by precipitation. But fine dusts and gases may reach the stratosphere where they remain for up to three years, being spread around the planet by high-level winds.

Mount Pinatubo erupted in the Philippines in June 1991, blasting 15 million tons of dust and gases high into the stratosphere. Within a year, the average temperature around the globe fell by 0.9°F, bringing harsh winters and cool summers to many parts of the world. Places lying directly beneath the narrow plume of dust and gases cooled first but after several months the volcanic sunscreen spread out, drifting toward the poles and cooling other areas. Some regions cooled

▼ **The eruption of Mount Pinatubo** in June 1991 threw fine dusts and gases, shown in yellow, high into the atmosphere (*top*). Two months later (*bottom*), winds had spread the material around the equator.

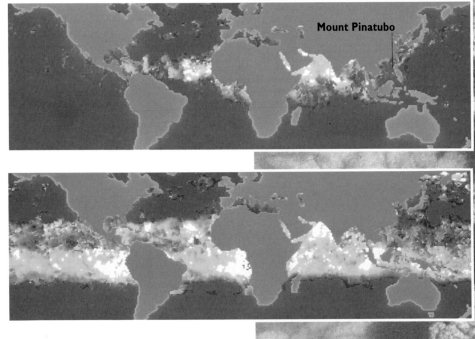

Mount Pinatubo

more than others and this upset the pattern of global winds and storm tracks, causing a cloudy, cool, and wet summer in some places that normally had a dry, warm summer.

A massive eruption at Tambora in Indonesia in 1815 produced "The Year Without a Summer" in 1816. Snow and frosts during that summer in Europe and North America killed many crops. Food shortages caused widespread famine and misery.

Bang or whimper?

Usually, the bigger the eruption, the greater its effect on climate. But this is not always the case. When Mount St. Helens in Washington State erupted in 1980 scientists expected it to chill the planet. Thick ash fell, covering crops and highways but the eruption came out sideways, so the dust and gases failed to reach the stratosphere and only places near the volcano experienced cooling.

The location of an erupting volcano is important. On reaching the stratosphere, dust and gases generally drift toward the nearest pole. Eruptions of volcanoes in Iceland, for example, mostly affect the northern continents, while volcanoes near the equator, such as Mount Pinatubo and Tambora, can affect both hemispheres.

◀ **Plumes of ash, smoke, dust,** and gases punch through the atmosphere from a volcanic eruption. As a result, the climate may cool for as long as three years.

SUNSPOTS

Dark spots have been observed on the face of the Sun for many centuries. They are called sunspots and the number of them changes over the years. Many sunspots mean that the Sun is hotter and more active. Few sunspots indicate a weaker, cooler Sun. The change from many sunspots to few sunspots and back again takes about 11 years.

Some scientists believe that droughts, rainfall, and cold winters in some parts of the Earth are influenced by this sunspot cycle. Not everyone agrees.

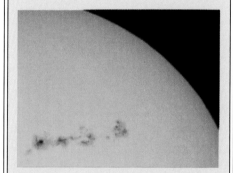

▼ **From around 1450 to 1850,** Europe suffered many bitterly cold winters. Frost Fairs were held on the thick ice of the Thames River in London. This "Little Ice Age" occurred during an unusually long period of very few sunspots.

El Niño

Oceans have a major effect on weather and climate. Cold seas chill the air above, so few clouds form. Warm seas, however, encourage moist air to rise and form deep rain clouds.

Sometimes the locations of warm and cold sea areas change. This happens in the Pacific Ocean every few years when the cold area in the eastern Pacific and the warm western area switch. The winds reverse, too, and this vast change, known as El Niño, plays havoc with the world's weather.

▲ **Destructive floods** and mudslides can result from the heavy rains that El Niño brings to Peru, Ecuador, and California (pictured).

◄ **El Niño brings** severe drought to Australia. Withering heatwaves, blinding dust storms, and wildfires add to the problems caused by the lack of water.

Weather upheaval

The name El Niño means "boy" in Spanish. It describes the shifting of warm water and a wet climate from the western to the eastern Pacific and the reversal of the trade winds. This leads to major changes in rainfall, temperature, and storm tracks around the world. Australia, Borneo, and Indonesia on the western side of the Pacific suffer drought. The Asian monsoon (*see pages 36–37*) is weakened and India and China receive less rain. Even far away in the Sahel, southern Africa, and Brazil, droughts occur or worsen.

During the early stage of El Niño, typhoons form in the central Pacific, sometimes striking islands such as Hawaii and Tahiti with devastating force. At the height of El Niño, torrential rains over northern Peru, Ecuador, and California lead to damaging floods and mudslides.

El Niño is brought about by very slow waves that slop back and forth across the Pacific Ocean basin, taking one to two years to do so. Each El Niño follows a similar pattern, but differs in strength.

La Niña

When the eastern Pacific has cold seas and a dry climate, it is known as La Niña (which means "girl" in Spanish). La Niña was long thought to be the normal situation, with El Niño being an unusual event, occurring every three to seven years. Today El Niño is returning more often and is staying longer.

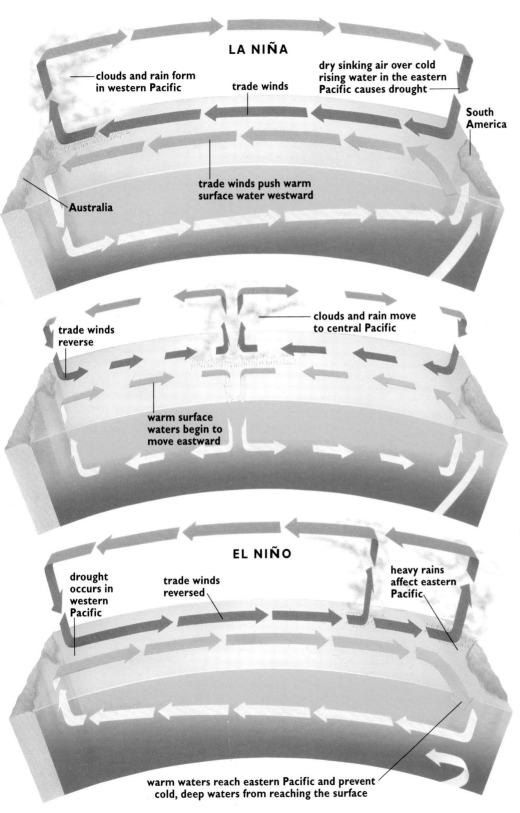

LA NIÑA

clouds and rain form in western Pacific

trade winds

dry sinking air over cold rising water in the eastern Pacific causes drought

South America

Australia

trade winds push warm surface water westward

trade winds reverse

clouds and rain move to central Pacific

warm surface waters begin to move eastward

EL NIÑO

drought occurs in western Pacific

trade winds reversed

heavy rains affect eastern Pacific

warm waters reach eastern Pacific and prevent cold, deep waters from reaching the surface

Wildlife disaster

During La Niña, cold water off the eastern Pacific coast rises and brings nutrients to the surface, which support abundant food for fish. El Niño prevents this from happening as warm surface waters spread eastward. Fish die or swim south to cooler Chilean waters. Pelicans and other seabirds as well as fur seals and sea lions that feed on the fish starve. Peruvian fisheries, which have thrived on huge catches of anchovies and sardines, go out of business during El Niño.

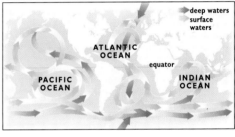

deep waters
surface waters

ATLANTIC OCEAN

equator

PACIFIC OCEAN

INDIAN OCEAN

▲ **Cold water sinks to great depths** in the North Atlantic, driving a worldwide pattern of deep-sea and surface currents. Water takes between 500 and 2,000 years to complete one trip around the world. Climate is affected if this oceanic conveyor belt changes even slightly.

◄ **During La Niña** (*top*), the waters of the western Pacific are warm, encouraging a wet climate. In the eastern Pacific, cold water rising from the deep ocean chills the air above and brings drought. Every few years, warm surface waters move eastward and the winds reverse (*middle*). During El Niño (*bottom*) the warm waters bring a wet climate to the eastern Pacific. Heavy rains affect the South American coast. Severe drought occurs in Australia.

Polluting the atmosphere

Human activities release vast amounts of pollution every day. Exhaust fumes from 700 million motor vehicles worldwide, together with plumes of smoke and gases from industry, send a huge variety of pollutants into the atmosphere.

The thick layer of pollution released by a large city can create a choking, brownish-gray smog, especially when there is no wind to blow the pollutants away. Acid gases from factory chimneys and motor vehicles mix with fog, rain, and snow, and result in acid rain, which damages buildings and forests and kills fish. Some pollutants escape into the stratosphere and destroy the natural ozone gas that protects animals and plants from the Sun's harmful ultraviolet rays.

▲ **Acid rain damages crops,** kills fish by poisoning lakes, and makes stone crumble. Trees lose their leaves and eventually die.

some pollutants are carried long distances by winds

some pollutants fall near their source

sulfur dioxide and nitrogen oxides are produced by industry, motor vehicles, and homes

sources of pollution: burning coal, wood, and oil

City smogs

Dangerous smogs form in towns and cities especially on calm days and when the sinking air of a high pressure zone stops pollution from escaping upward.

In many cities, motor vehicles cause more pollution than factories and they are rapidly increasing in number. When you wake up each morning, 30,000 new vehicles will have been added to the world total.

Because pollutants in the air can affect breathing, warnings of smogs may be given out. These are aimed particularly at people with asthma and other breathing problems—as many as one in seven children in some countries. "Don't jog in the smog" and similar slogans advise people to avoid outdoor exercise because this leads to breathing in more pollutants. Sometimes, the advice is not to go out at all! Cars may be banned from city centers and factories shut down to reduce pollution.

▲ **Gases, mostly sulfur dioxide** and nitrogen oxides, are released from burning oil and coal and become dissolved in rain, snow, and fog. This acid precipitation, or acid rain, is harmful to animals and plants.

Acid rain

Gases mix with moisture in the air and can add acid to precipitation, making it like lemon juice. When the precipitation falls over forests, it washes nutrients from the soil and from leaves and needles, which turn yellow and die. Acid rain releases aluminum, which is present in the soil but is usually "locked up." This weakens tree roots, making the trees more easily damaged by pests, frosts, and strong winds.

acid gases mix with water in air, making precipitation acidic

acid rain discolors and kills leaves and needles

tree roots are damaged by the aluminum released from the soil

acid waters damage fish eggs and aluminum released into lakes makes it difficult for fish to breathe

▼ **City smogs** like this one in Los Angeles cause breathing problems, especially among children, whose airways are narrower.

THE OZONE HOLE

Each year between September and November, the layer of ozone gas in the stratosphere above the Antarctic thins dramatically. Pollutants, especially chemicals called chlorofluorocarbons (or CFCs), destroy the ozone when the Sun returns after the long, dark polar winter. An ozone hole the size of the United States and as deep as Mount Everest is created. CFCs are now banned but those that already exist will remain in the atmosphere for tens of years. CFCs were once used in aerosol sprays, in plastic foam used in food packaging, and as the coolant in refrigerators and air conditioning units in cars.

Ozone filters out most of the harmful ultraviolet rays from the Sun, which would otherwise damage eyes, cause sunburn, lead to skin cancers, and restrict the growth of crops.

▲ **A large ozone hole** (shown at center in dark blue and gray) develops over Antarctica for a few months each year.

▶ **Hats and sunblock creams** give some protection against the painful sunburn that the Sun's ultraviolet rays can cause.

The aluminum is washed into lakes and causes fish to overproduce a sticky mucus which clogs their gills and makes breathing difficult. The acid waters also damage fish eggs. Eventually all the fish die.

Reducing pollution

Many countries are now trying to reduce pollution by using cleaner fuels in industry and motor vehicles. The number of cars powered by natural gas and electric batteries is increasing, and people are being encouraged to use their cars less, especially for short journeys. Improvements in bus, subway, and rail services are also needed because public transport causes less pollution per passenger than cars.

The greenhouse effect

As the world's population grows, more and more coal, oil, gas (all known as fossil fuels), and wood are being burned to produce the energy needed for heating, cooking, transporting, building, and making the goods people need and the luxuries they want. Some gases released by the burning of these fuels are called greenhouse gases because they act like the glass of a greenhouse: they let sunlight through but trap heat given off from the Earth's surface. As a result, they warm the planet. The more gases there are in the atmosphere, the greater the warming will be.

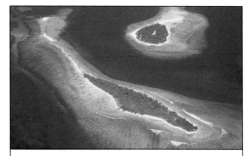

The sea level may rise by between 6 and 18 inches by the year 2050. Low-lying islands, coastal cities, and farmland will need to be protected. Mangrove forests and coral reefs may be destroyed.

Global warming

During the past 140 years, the world climate has warmed by 1.1°F. About half of the greenhouse gases that enter the atmosphere each year remain there. This means that the amount of greenhouse gases is increasing and the planet is expected to continue to warm.

▼ **Industry burns fossil fuels,** releasing greenhouse gases into the atmosphere which are warming the Earth.

Greenhouse gases

Without the blanketing effect of greenhouse gases, the Earth's average temperature would be around 0°F instead of 59°F, so they are vital for our survival. What is worrying scientists, however, is the extra warming that the increasing amounts of greenhouse gases will produce during the next century.

Carbon dioxide is the gas responsible for about half the global warming during the past 140 years. It is released when fossil fuels are burned and when tropical forests are burned to clear land for farms and towns. Fewer trees also means that less carbon dioxide is removed from the air, since trees take in carbon dioxide as they grow. Other important greenhouse gases are CFCs, methane, nitrous oxide, ozone, and water vapor.

methane from animals' digestion, rice paddy fields, swamps, rotting vegetation, and gas pipes

sunlight reflected off clouds

nitrous oxide from car exhausts and fertilizers spread on fields

sunlight reflected off Earth's surface

Some regions will become drier, others wetter. Drier soils will suit different crops, need to be irrigated, and require farming methods that hold soil moisture. Pests and diseases may thrive in warmer conditions.

Forests are slow growing. Commercial forests need 50 years before the trees can be harvested. In that time, climate will change and the trees may not be able to adapt.

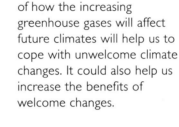

As climate warms, animals, birds, and plants that need cool conditions will have to migrate poleward or adapt. Some plants and animals may thrive and multiply; others will die.

FUTURE CLIMATE?

Industrialized nations have agreed not to increase the amount of greenhouse gases they release into the atmosphere. But this will only slow the warming. Drastic reductions in greenhouse gases are needed if the warming is to be stopped. Scientists estimate that the Earth's climate may warm by between 3 and 6°F during the next century, making it hotter than it has been for 135,000 years.

Increases in the average temperature of the Earth's climate

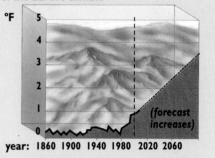

°F 5 4 3 2 1 0 (forecast increases)

year: 1860 1900 1940 1980 2020 2060

some heat escapes into space

sunlight

CFCs from aerosols, foam plastics, and old, destroyed refrigerators

▲ Accurate forecasting
of how the increasing greenhouse gases will affect future climates will help us to cope with unwelcome climate changes. It could also help us increase the benefits of welcome changes.

heat trapped by greenhouse gases, warming the atmosphere (greenhouse gases include carbon dioxide, methane, CFCs, nitrous oxide, water vapor, and ozone)

carbon dioxide from burning forests and grassland

carbon dioxide from burning fossil fuels

heat from the Earth

◀ The Earth's temperature
is controlled by the amount of sunlight reaching the ground and the amount of heat escaping from the Earth. The build-up of greenhouse gases from various sources is changing this energy balance by preventing some heat from escaping.

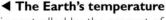

methane from rotting refuse

Weather and climate changes

The Earth will not warm evenly. The polar regions will warm more than the tropics because as snow and ice melt, sunlight warms the ground rather than being reflected back into space. Patterns of winds, storms, and ocean currents will change as they will need to carry less warmth toward the poles. This may result in some areas getting more rain and others less. Warmer oceans may lead to more thunderstorms and even tornadoes. Ocean areas that are warmer than the 80°F needed for hurricanes to form will become larger, and coastal cities will be threatened by higher sea levels.

Glossary

If you cannot find the word you are looking for here, try looking in the index. Most words having to do with weather are explained in the text.

air mass   A large body of air hundreds to thousands of miles across that has similar temperature and humidity throughout. The edge of an air mass is marked by a **front**.

anticyclone   (*see* **high**)

anvil (cloud)   The shape formed by cirrus clouds at the top of a cumulonimbus when they are stretched out by high-level winds. The shape is similar to that of a blacksmith's anvil and often marks the top of the troposphere.

atmosphere   The envelope of gases that surrounds the Earth and which contains the oxygen we breathe.

avalanche   A rapid movement of snow down a steep slope.

blizzard   Severe weather conditions of heavy snow, strong winds, low temperature, and poor visibility.

cirrus   A high, thin, white feathery cloud.

climate   The typical pattern of daily and seasonal weather of a place.

condense   To change from a gas into a liquid, such as from water vapor into water droplets, usually by cooling.

Coriolis force   A force resulting from the Earth's rotation, which causes a wind to appear to turn to the right of its path in the northern hemisphere and to the left in the southern hemisphere. The Coriolis force is strongest at the poles and weakens toward the equator where it disappears altogether.

cumulus   A dome- or tower-shaped white cloud with a flat base and a bulging top.

cyclone   (*see* **low** *and* **tropical cyclone**)

depression   (*see* **frontal depression**)

downburst   A powerful downward surge of air from the base of a severe thunderstorm. A **microburst** is a violent downburst less than 2½ miles across.

drought   A period of unusual dryness causing water shortages.

equator   An imaginary line drawn around the Earth, halfway between the poles.

evaporate   To change from a liquid into a gas, such as from water droplets into water vapor, usually by warming.

fog   A cloud with its base touching the ground.

forecast (weather)   A statement of how the weather is likely to change in the future in a particular place.

front   The leading edge of a moving body of air (an air mass). Cold air lies behind a cold front, warm air behind a warm front.

frontal depression   A low in which air masses of different temperatures and humidities meet and rotate slowly around a central point. It brings unsettled, changeable weather.

greenhouse effect   The warming of the Earth caused by particular gases in the atmosphere, called **greenhouse gases**. These allow sunlight to reach the Earth's surface, but trap heat given off by the ground.

hemisphere   Half the Earth divided by the equator. The northern hemisphere lies to the north of the equator; the southern hemisphere lies to the south.

high   A region of high air pressure around which winds blow clockwise in the northern hemisphere and counterclockwise in the southern hemisphere. It is also called an **anticyclone**. A high brings dry weather.

humidity   The amount of water vapor present in the air. Dry air has a low humidity, moist air a high humidity.

hurricane   (*see* **tropical cyclone**)

isobar   A line on a weather map joining places that have the same air pressure.

jet stream   A narrow band of strong winds in the upper levels of the troposphere.

latitude   A measurement in degrees that is used to describe how far a location is to the north or south of the equator: the equator is 0°, the North Pole is 90°N and the South Pole 90°S. The **middle latitudes** lie between 30° and 60° latitude in each hemisphere.

low   An area of low pressure around which the winds blow counterclockwise in the northern hemisphere and clockwise in the southern hemisphere. It is also called a **cyclone**. A low often brings cloud, precipitation, and windy weather.

meteorologist   A person who studies **meteorology**, which is the study of how the atmosphere creates weather and climate.

microburst   (*see* **downburst**)

middle latitudes   (*see* **latitude**)

monsoon   The seasonal reversal of winds that results in a wet summer season and a dry winter season in several regions of the world. It is strongest in southern Asia, northern Australia, and western Africa.

occlude   When referring to a frontal depression, this describes the point at which a cold front catches up with a warm front and lifts the warm air away from the ground to produce an **occluded front**. This happens toward the end of the life of a frontal depression as it starts to dissolve.

ozone   An almost colorless gas. The largest amounts are found between 9 and 25 miles high in the stratosphere, where it absorbs some of the harmful ultraviolet rays from the Sun. This is called the **ozone layer**. Ozone also forms in polluted air near the ground, being part of summertime smogs.

polar   Something from or to do with the Earth's poles. A polar region, for example, is the region around a pole. Polar air comes from a polar region.

polar front   The boundary (or front) between warm tropical air and cold polar air which encircles each hemisphere in the middle latitudes. Frontal depressions form and travel along it.

precipitation   Any form of water—whether liquid, such as rain, or solid, such as snow—that falls from the atmosphere and reaches the ground.

pressure (air)   The force of the air pressing down on the Earth's surface.

ridge   A lengthened area of high air pressure.

stratosphere   The layer of the atmosphere that lies above the troposphere.

stratus   A low, ragged, layer-shaped gray cloud often producing drizzle.

temperate   The name given to the typical climate conditions of the middle latitudes, which are influenced from time to time by both tropical and polar air masses.

temperature   A measure of the hotness or coldness of a substance. Temperature is measured using a thermometer.

thermal   A rising current (or large bubble) of air produced by intense heating of the ground. Cumulus clouds show the presence of thermals.

thunderstorm   A cumulonimbus cloud producing lightning and thunder.

tornado   An intense, rotating column of air that extends downward from the base of a cloud, usually a cumulonimbus, in the shape of a funnel, column, or rope and which touches the ground. A funnel cloud is a tornado that does not reach the ground.

trade winds   Steady winds in the tropics blowing from the northeast in the northern hemisphere and from the southeast in the southern hemisphere.

Tropic of Cancer/Tropic of Capricorn
Imaginary lines drawn around each hemisphere, the Tropic of Cancer being at latitude 23.5°N and the Tropic of Capricorn at 23.5°S. They mark the limits of the area of the Earth's surface in which the Sun can be directly overhead.

tropical   Something from or to do with the tropics. For example, tropical air comes from the tropics.

tropical cyclone
A violent tropical storm with winds blowing at 74 mph or more around a low pressure center. It is known as a **hurricane** in the Atlantic Ocean and eastern Pacific Ocean, a **typhoon** in the northwestern Pacific, and a **cyclone** in the Indian Ocean and around Australia.

tropics   The region of the Earth between the tropics of Cancer and Capricorn.

troposphere   The layer of the atmosphere that lies closest to the Earth. It contains most of the weather.

trough   A lengthened area of low air pressure.

typhoon   (*see* **tropical cyclone**)

vapor (**water**)   Water in the form of an invisible gas.

visibility   The greatest distance an observer can see and identify certain objects.

waterspout   A rotating column of air that extends downward to touch a water surface from the base of a cumulus or cumulonimbus cloud. It is usually less violent than a tornado.

weather   What is happening in the atmosphere outside at any particular time and place in terms of clouds, humidity, precipitation, sunshine, temperature, visibility, wind, and air pressure.

wind   Moving air. Wind direction is the direction from which the wind is blowing.

Index

Note: Page numbers in **bold** give main references; page numbers in *italic* refer to information given only in boxes or in picture captions or labels.

acid rain 62, 63
air mass 10, *11*, 18
air pressure 14–15, *45*, 47
altocumulus clouds *11*
altocumulus lenticularis clouds *52*
altostratus clouds 10, *11*, 19
anticyclones 15
  see also highs
asteroids 55
atmosphere 8, 9, 10
  and global warming 64–65
  layers in 8, *9*
  moving heat 16, 32
  pollution of 62–63
  water in 10, *13*
aurorae *9*
avalanches *40*, 41

barometer *see* weather instruments
Beaufort Scale 15
blizzards 40
blocking highs 42
blue skies *see* sky, colors of

carbon dioxide 64, *65*
CFCs *63*, 64
chinook 17
chlorofluorocarbons *see* CFCs
cirrocumulus clouds *11*
cirrostratus clouds *11*, 19
cirrus clouds *11*, 19, 51, 53
city climates 20
  see also smogs
clear skies 15, *18*, 42, 52
climate change 54–65
  adapting to *65*
  evidence of past 54–55

climates 20–21
clouds **10–11**, 15, *17*, 18–19, 45, 52, 53, 60
  and precipitation 12–13
  depth of 53
  formation of **10–11**
  seeding 13
  types of *11*
  see also names of clouds
cold forest climate *21*
condensation level *10*
condensation trails *see* contrails
contrails 53
Coriolis force 15, *16*, 34
cumulonimbus clouds 10, *11*, *13*, 16, *18*, *19*, 28
cumulus clouds 10, *11*, *18*, 22, 28, 52, 53
cyclone shelters 35
cyclones *see* hurricanes

desert climate *21*
dew *12*
dinosaurs, extinction of 55
downbursts **26–27**, 49
drizzle *12*, 13, 18
drought **42–43**, 55, 59, 60, 61
dry grassland climate *21*
dust devils 29
dust storms 42, *43*, 60

Earth's axis *9*, 56, 57
Earth's orbit 56, 57
El Niño **60–61**

floods 32, 36, 37, 60
  flash 22, **38–39**
  prevention of *39*
  river *37*
fog 10, *17*, 49, 62
föhn *17*
fossil fuels 64
fossils 54, 55
freezing rain 12
frontal depressions 15, 16, *18*, *19*, 42, 53
fronts 18–19
frost *12*, 59
Frost Fairs *59*
funnel cloud 28

gales 15, 40, 49
glacial conditions *see* ice ages
glaciers 56, 57
glaze *see* freezing rain
global circulation *16*
global cooling 55, 58
  see also ice ages
global warming 64–65
greenhouse effect 64–65
greenhouse gases 64, *65*

hail 13, 26, **27**, 28
halos *50*, 51
highs 14–15
  and pollution 62
  see also blocking highs
humidity *45*
hurricanes 15, 16, **32–35**
  and flash floods 38
  and global warming 65
  damage scale 35
  Hurricane Andrew 35
  Hurricane Beulah 32
  Hurricane Gilbert 32, 35
  naming of 35

ice ages 56–57
ice cores 55
ice crystals 10, 12, 13, 22, 50, 51, 53
ice sheets 55, 56, 57
  bodies preserved in 57
instrument shelter *see* weather instruments
interglacial conditions *see* ice ages
isobars 14, 47

jet *9*, 22, 23
jet stream 16, **18**, *19*, 42
Jimsphere *45*

La Niña **60–61**
land breeze *17*
lenticular clouds *see* altocumulus lenticularis
light *see* sunlight
lightning 22, **23**, 24–25
lows 14–15, 42

mamma (mammatus) cloud 28, 53
Mediterranean climate 20

microbursts *see* downbursts
monsoon 20, **36–37**, 60
mountain breeze *17*
mountain climate *21*
mountains, effect on weather *11*, 17, 20, 52
Mount Pinatubo 58, 59
Mount St. Helens 59

National Hurricane Center, Miami 35
nimbostratus clouds 10, *11*, 18, *19*

ocean currents 20, 55, *61*, 65
ozone gas 62, *63*, 64
ozone hole 63

polar air 18, 52
polar climate *21*
polar front 16, 18
pollution 50, **62–63**
precipitation **12–13**, 17, *21*, 45
  and climate 20
  see also dew, drizzle, freezing rain, frost, hail, rain, showers, sleet, snow

rain 12, 13, 15, 16, *17*, 18, *19*, 26, 49, 52, 53, 60, *61*, 62
  see also city climates, drought, floods, monsoon
rainbows 50–51
rain gauge *see* weather instruments
red skies *see* sky, colors of
ridges 19

salting roads 49
sanding roads 49
Santa Ana *17*
satellites 8, 34, **44**, *45*, 46
sea breeze *17*, 52
sea ice 56, 57
sea levels
  falling 57
  rising 64, 65
showers *11*, 12, 53
sky, colors of 50, *51*
sleet 13
smog 62, 63

snow 12, 13, *17*, 26, 40–41, 49, 53, 56, 59, 62
snowflakes 12
sprite *9*, 22, 23
storm chasers 29
storm surge 33, 35
stratocumulus clouds 11, *19*
stratosphere 9, 58, 59, 62, 63
stratus clouds 10, *11*, *19*
sun dogs 51
sunlight/sunshine *9*, 16, *45*
 colors in 50
 effects in sky 50–51
 ultraviolet rays 62, 63
 see also global cooling, global warming
sun pillars 51
sunspots *59*

Tambora 59
temperate climate 20, *21*
temperature 8, 10, 40, 44, 45, 49, 60, 64
 and climate *20*
 see also global cooling, global warming
thermal *10*, *11*
thermograph *see* weather instruments
thermometers *see* weather instruments
thunder 23
thunderstorms 9, 20, **22**, 23, 53
 and downbursts 26
 and flash floods 38, *39*
 and global warming 65
 and hail 26, 27
 and hurricanes 32, 33, 34
 and tornadoes 28
 and waterspouts 30
Tornado Alley *31*
tornadoes **28–31**, 32
 and global warming 65
Totable Tornado Observatory (TOTO) 31
trade winds 16, 33, 60, *61*
tree rings 54, *55*
tropical air 18
tropical climates *20*
tropical cyclones *see* hurricanes
tropics 16, 20

troposphere **8**, **9**, 10, *11*, 58
troughs 18, *19*
tundra climate *21*
typhoons *see* hurricanes

valley breeze *17*
virga 53
visibility *45*, 52
volcanic eruptions 55, **58–59**

water cycle *13*
water droplets 10, 12, *13*, 22, 50
waterspout *30*, *31*
water vapor 10, *11*, 12, *13*, 64
weather agencies *49*
weather balloons 44, *45*
weather forecasts 19, **44–49**
 and computers 46, 49
 using forecasts **48–49**, 65
 wrong forecasts 49
 see also weather maps
weather instruments 34, 45
weather layer *see* troposphere
weather maps 14–15, 47
weather observations 45
weather sayings 50, 52
weather stations 44
weather symbols 47
winds **14–17**, 47, 49, 52, 55, 59,
 direction 16, *45*, 47
 local 17
 and pollution 62
 speeds 14, 15, *45*, 47
 strength 15, *45*, 47 *see also* Beaufort Scale
wind shear 26

Year Without a Summer, The 59

PICTURE ACKNOWLEDGMENTS

ARTWORK

Peter Sarson/Richard Chasemore, 8–9 (top), 9 (top right), 10, 11 (center), 12 (top left), 12–13 (center), 14–15, 16, 17 (top left), 22–23, 25, 26–27, 28, 30–31 (top), 31 (bottom right), 33 (bottom right), 34–35 (top), 44–45 (center), 50, 51 (bottom right), 52 (top left), 53 (center), 56, 62 (top), 62 (bottom), 63 (center); **Michael Welply**, 45 (center right); **Rob Jakeway**, 8–9 (bottom), 11 (top right), 18–19, 32–33, 51 (top right); **Gary Hincks**, 13 (bottom right), 17 (bottom right), 21 (bottom right), 36–37 (bottom), 38–39, 41 (bottom), 42–43 (center), 48–49 (center), 61, 62–63 (center), 64–65 (center); **Richard Bonson**, 54–55 (center); **Eugene Fleury**, 20–21, 31 (top right), 34 (bottom left), 43 (bottom right), 46–47 (center), 57 (bottom left), 65 (center right); **Michael Roffe**, 46–47 (insets), 48–49 (insets), 64–65 (insets); **Michael Woods**, 30 (top right), 30 (bottom right); **Mark Franklin**, 47 (right), black icons; **Steve Kirk**, 30 (center right); **Alan Male**, 30 (top left).

PHOTOGRAPHS

Endpapers: John Eastcott/Planet Earth Pictures; 4 John Shaw/Tom Stack and Associates; 7 John Barlow; 8 European Space Agency/Science Photo Library; 10/11 John Eastcott and Yva Momattuk/Science Photo Library; 12 Zefa Pictures; 14 Alastair Black/Tony Stone Images; 17 Colin Prior/Tony Stone Images; 18 John R. Bracegirdle/Planet Earth Pictures; 20 (top) Trevor Wood/Tony Stone Images, 20 (center) K. Gillham/Robert Harding Picture Library, 20 (bottom) Tony Morrison/South American Pictures; 21 (top left) Zefa Pictures, 21 (top center) Ben Osborne/Tony Stone Images, 21 (top right)Philippe Henry/Oxford Scientific Pictures, 21 (bottom and center right) Zefa Pictures; 24 (center) H. Bird/Frank Lane Picture Agency; 25 John Autrey/Science Photo Library; 26 James King-Holmes/Science Photo Library; 27 (top) Mary Evans Picture Library, 27 (bottom left) Suddeutscher Verlag, 27 (bottom right) Frank Oberle/Tony Stone Images; 28 Rex Features; 29 Robert Harding Picture Library; 30 Richard Chesher/Planet Earth Pictures; 33 Flip Schulke/Planet Earth Pictures;34 NASA/Science Photo Library; 35 (top) Robert Holland/The Image Bank, 35 (bottom) Heldur Jaan Netocny; 36 (top) Steve McCurry/Magnum Photos, 36 (bottom) Images Colour Library; 37 Steve McCurry; 38 Christian Borderie/Reuters/Popperfoto; 40 (top) E. Dragesco/Ardea, 40-41 (background) Zefa Pictures; 42 Jean-Paul Ferrero/Ardea; 43 Zefa Pictures; 44 NOAA; 45 David Parker/Science Photo Library; 48 Tony Stone Images; 49 Reuter/Popperfoto; 50 (bottom) Mark Newman/Frank Lane Picture Agency, 50-51 (background) Images Colour Library, 50-51 (top) Images Colour Library; 52 (bottom) Martyn Chillmaid/Oxford Scientific Films, 52-53 (background) Adam Jones/Planet Earth Pictures; 53 (top) Zefa Pictures, 53 (bottom) David Hoadley/Frank Lane Picture Agency; 54 Vaughan Fleming/Science Photo Library, 54-55 John Barlow; 55 (top) Kunsthistorisches Museum, Vienna/AKG London, 55 (bottom) J.G.Paren/Science Photo Library; 56-57 (background) Zefa Pictures, 56-57 (top) Zefa Pictures; 57 (bottom) Popperfoto; 58 Robert M.Carey, NOAA/Science Photo Library; 58-59 J. Lynch/Frank Lane Picture Agency; 59 (top) John Sanford/Science Photo Library, 59 (bottom) Guildhall Library, London/The Bridgeman Art Library; 60 (top) Frank Spooner Pictures, 60 (bottom) David Austen/Frank Spooner Pictures; 62 Richard Packwood/Oxford Scientific Films; 63 (top) NOAA/Science Photo Library, 63 (bottom left) Deborah Davis/Tony Stone Images, 63 (center right) Zefa Pictures; 64 (bottom left) Jeremy Walker/Tony Stone Images, 64 (top right) Silvestris/Frank Lane Picture Agency, 64 (bottom right) Zefa Pictures; 65 (top left) David Carriere/Tony Stone Images, 65 (center top) Yann Arlhus Bertrand/Ardea.

If the publishers have unwittingly infringed copyright in any illustration reproduced, they would pay an appropriate fee on being satisfied to the owner's title.